Lecture Notes in Mobility

Series editor

Gereon Meyer, VDI/VDE Innovation und Technik GmbH, Berlin, Germany

More information about this series at http://www.springer.com/series/11573

Michael Behrisch · Melanie Weber
Editors

Simulating Urban Traffic Scenarios

3rd SUMO Conference 2015 Berlin, Germany

Editors
Michael Behrisch
Institute of Transportation Systems
German Aerospace Center (DLR)
Berlin
Germany

Melanie Weber
Institute of Transportation Systems
German Aerospace Center (DLR)
Berlin
Germany

ISSN 2196-5544   ISSN 2196-5552   (electronic)
Lecture Notes in Mobility
ISBN 978-3-319-33614-5   ISBN 978-3-319-33616-9   (eBook)
https://doi.org/10.1007/978-3-319-33616-9

Library of Congress Control Number: 2018941981

Printed on acid-free paper

This Springer imprint is published by the registered company Springer International Publishing AG part of Springer Nature
The registered company address is: Gewerbestrasse 11, 6330 Cham, Switzerland

Preface

Urban traffic is a major challenge to the traffic simulation community and traffic engineering in general. The heterogeneity of traffic modes is still increasing with electric bikes and multiple sharing modes calling for new models and increased complexity for existing simulators. On the other hand, traffic engineers are supported by readily available tools, such as the open-source package Simulation of Urban Mobility (SUMO), which can build a working prototype of a multimodal simulation scenario in virtually no time based on open data services like OpenStreetMap. Furthermore demand data, either stemming from public transport data or from the evaluation of people's day plans, can now be collected with the support of a networked and networking population which enables the detailed estimation of the effects of engineering measurements as well as emerging new technologies through the means of individual (microscopic) traffic simulation. This simulation of every single actor allows the integration of behavioral data which can interface with the existing models to gather new insights into the social dynamics of traffic as well.

This volume contains extended versions of papers presented at the third SUMO User Conference (SUMO2015), which was held during May 7–8, 2015, in Berlin-Adlershof, Germany. SUMO is a well-established microscopic traffic simulation suite which has been available since 2001 and provides a wide range of traffic planning and simulation tools. The conference proceedings give a good overview of the applicability and usefulness of simulation tools like SUMO ranging from the in-house logistics applications and traffic signals to the impact of floods on the traffic of complete cities. Another aspect of the tool suite, its universal extensibility due to the availability of the source code, is reflected in contributions covering network modification tools and workflow improvements to develop microscopic traffic simulation scenarios.

Several articles give outlines of detailed aspects of public transport modeling and inner city vehicle networks when setting up a simulation with SUMO as well as an overview about the application of the tool in large-scale scenarios or for traffic light evaluation. The conference series's aim is bringing together the large international user community and exchanging experience in using SUMO, while presenting results or solutions obtained using the software. This collection should inspire you to try your next project with the SUMO suite as well or to find new applications in your existing environment.

Berlin, Germany   Michael Behrisch
December 2015   Melanie Weber

SUMO2015 Organization

SUMO2015 was organized by the Institute of Transportations Systems, German Aerospace Center, Berlin.

International Scientific Committee

Constantinos Antoniou, National Technical University of Athens, Greece
Ana Bazzan, Universidade Federal do Rio Grande do Sul, Brasil
Hilmi Berk, Istanbul Technical University, Turkey
Robbin Blokpoel, Imtech Traffic & Infra, The Netherlands
Quentin Bragrad, University College Dublin, Ireland
Christine Buisson, Laboratoire Ingénierie Circulation Transport, France
Winnie Daamen, Delft University of Technology, The Netherlands
Jakob Erdmann, German Aerospace Center, Germany
Jorge E. Espinosa Oviedo, Universidad Nacional de Colombia, Colombia
Gunnar Flötteröd, KTH Royal Institute of Technology, Sweden
Ellen Grumert, Linköping University, Sweden
Florian Hagenauer, University of Paderborn, Germany
Marek Heinrich, German Aerospace Center, Germany
Mario Krumnow, University of Technology Dresden, Germany
Vincenzo Punzo, University of Napoli, Italy
Rosaldo Rossetti, University of Porto, Portugal
Andreas Schadschneider, University of Cologne, Germany
Sebastian Schellenberg, Friedrich-Alexander University of Erlangen-Nürnberg, Germany
Jörg Schweizer, University of Bologna, Italy

Organization Committee

Michael Behrisch, German Aerospace Center, Germany
Melanie Weber, German Aerospace Center, Germany

Contents

Part I
Tools and Models

Application of the SCRUM Software Methodology for Extending Simulation of Urban MObility (SUMO) Tools

Andrés F. Acosta, Jorge E. Espinosa and Jairo Espinosa

Abstract This chapter explains the implementation of software tools for setting up multimodal simulation scenarios in the Simulation of Urban MObility (SUMO). This implementation has been supported by the SCRUM methodology, which is a methodology suitable for incremental software development. Implementation of these tools starts from the re-engineering of open-source tools in previous developments. Furthermore, the new components are described and tested in a simulation scenario in the city of Medellín.

1 Introduction

Road traffic simulations greatly support the decision-making processes regarding improvements in mobility for a city. Many advancements have been made in the last three decades regarding road traffic models and both commercial and open-source tools that simplify the simulation of realistic scenarios. In the open-source side, Simulation of Urban MObility (SUMO) is one of the most well-known and accepted, because it has a mature development process from 2002. Moreover, SUMO's vehicle model has been validated and it is widely accepted. SUMO also offers the possibility to import traffic networks from several sources, including Open Street Maps (OSMs).

However, one disadvantage of SUMO is that the preparation of simulation scenarios depends on the knowledge of some programming skills, which could be improved if supported by proper software tools. However, developing those tools from scratch could lead to a great effort, compared to the case of reusing available open-source packages.

J. E. Espinosa
Politécnico Colombiano Jaime Isaza Cadavid, Cra. 48 no. 7-151, Medellín, Colombia
e-mail: jeepinosa@elpoli.edu.co

A. F. Acosta (✉) · J. Espinosa
Universidad Nacional de Colombia, Cra. 80 no. 65-223, Medellín, Colombia
e-mail: afacostag@unal.edu.co

J. Espinosa
e-mail: jespinov@unal.edu.co

© Springer International Publishing AG, part of Springer Nature 2019
M. Behrisch and M. Weber (eds.), *Simulating Urban Traffic Scenarios*,
Lecture Notes in Mobility, https://doi.org/10.1007/978-3-319-33616-9_1

This chapter describes the use of several open-source tools, representing a continuation of the work made in [1, 2], in a re-engineering framework. The results of this work could integrate MATLAB® with SUMO through the TraCI4MATLAB package. Here, some improvements made to TraCI4MATLAB in the last year are mentioned. On the other hand, some extensions made to the open-source initiative Network Editor for SUMO (NES) [3] (an open-source SUMO Network Editor developed in Qt [4]) are explained, including a Model–View–Controller (MVC) [5] formulation that is being implemented to date. Improvements to NES include a road network import wizard and a traffic lights' editor. Last but not least, a re-implementation of the SUMOLib library for MATLAB® is described. This component allows the creation of traffic demand and network parsing for those users familiar to this programming language.

These tools can benefit the rapid creation of simulation scenarios in SUMO. Furthermore, they are tested in a simulation scenario in the city of Medellín, as a first step to achieve a larger scenario in the MOYCOT project [6].

Finally, these developments are carried out using the SCRUM agile software methodology, which resulted to be very adequate to the case of the re-engineering of open-source software because of the small developer team size, the closeness of developers and researchers (the customers), and the need to follow an incremental approach. It has been found that the SCRUM methodology can favor efficiency by focusing on functional software increments [7]. Although some disadvantages of agile methodologies have been mentioned in the literature, including a reduced attention to software engineering phases previous to implementation and the importance of taking into account the role of the customer [8], it has been found that this methodology can be easily combined with other software engineering principles and practices. In the case of the MOYCOT project, it was easy to incorporate some "best practices" related to the object-oriented re-engineering patterns [9, 10], as explained in [1, p. 4].

This chapter is organized, as follows: Sect. 2 explains the main characteristics of the SCRUM methodology and its incorporation in the MOYCOT project. Section 3 describes the new features implemented for TraCI4MATLAB and introduces SUMOLib4MATLAB and two improvements to NES for network importing and traffic lights' editing. Section 4 tests the aforementioned tools in a simulation scenario in Medellín. Finally, Sect. 5 presents conclusions.

2 Scrum

SCRUM is a methodology for agile software development able to deal with complex software projects delivering functional products in small iterations, called increments or Sprints, in the SCRUM jargon. SCRUM is highly adaptable and involves customers in regular meetings (SCRUM meetings) making it suitable to work when the requirements are unclear. Requirements are organized and prioritized in a list known as the *Product Backlog,* facilitating the focus on the most relevant functionality. Additionally, SCRUM defines some roles performed by the developer team and the customers, described as follows:

- The *SCRUM Master*. The person exerting this role is in charge of coordinating the interaction among the developer team and the customers, in the *SCRUM Meetings*
- The *Developer Team*. Their responsibility consists on implementing and delivering the software increments. This role acts by self-organizing the team oriented to the achievement of requirements, as explained in [11].
- The *Product Owner.* They represent the stakeholders that can provide feedback in the *SCRUM Meetings*, while in the stage of requirements specification are implemented.

As explained before, the responsibility of people involved in these roles includes the building of the *Product Backlog*, tracking the correct completion of the desired functionalities. Also, they must plan the *Sprints* and the requirement to be fulfilled. It is important to note that a more detailed monitoring of the time required for each requirement is made through the *Burning Charts* which is a representation of the remaining tasks necessary to fulfill a requirement against the time spent on them. Finally, SCRUM defines the *Daily SCRUMS,* which are meetings with an average duration of 30 min, where each developer can inform the team about the completion of his/her responsibilities and the problems found.

3 Application of the SCRUM Methodology in the MOYCOT Project

Modeling and Control of Urban Traffic in the city of Medellín (MOYCOT) is a research project whose main goals are to characterize multimodal traffic such as vehicles, pedestrians, bicycles, and BRT in the city of Medellín (Latin America), in order to propose optimal traffic light control strategies, thus reducing traffic congestion as much as possible. The SUMO simulator was selected after a literature and benchmark review because of its ability to be extended to incorporate traffic models and control. The MOYCOT project organized an international seminar [12], where stakeholders from the government and academics expressed their interest on having software tools for rapid setting up of road traffic simulation scenarios in the city. This seminar yielded some desired functionalities of a software package, including interactive graphical editing of road networks and traffic demand definition from sources such as O/D matrices. Also, the interaction in real time with external applications was formulated. For these reasons, later, it was concluded that SUMO was the right tool to use and extend, due, among others, to its great popularity and support. A short study in the MOYCOT project concluded that it was more appropriate to extend the existing software than developing a new simulator. In this regard, the concept of software re-engineering brings the best practices to follow in order to extend unknown software packages. This represented a considerable gain in effort.

The MOYCOT project identified three important needs:

- A flexible and expandable graphical editor. It is important to clarify that SUMO already has a graphical network editing tool, namely NETEDIT but unfortunately, to date, it is not open. Moreover, it allows to interact with external applications through the TraCI API [13].
- Means to connect with MATLAB®, since the research team involved in the MOY-COT project is experienced on the MATLAB® language, especially in the area of traffic control. Thus, it was concluded that the team could take advantage of that experience. Therefore, the first item in the product backlog was defined to be an implementation of the TraCI API for MATLAB®, called TraCI4MATLAB. After that, by identifying the Python library SUMOLib, which is part of the SUMO tools, the team planned to implement this component in MATLAB® and extend it to support the graphical editing functionalities, thus corresponding to the second item in the product backlog.

The development of this backlog evolved as requirements was completed, as described below.

3.1   Sprint No. 1: TraCI4MATLAB

As described previously, the first requirement in the product backlog was to develop TraCI4MATLAB, which was associated with this Sprint of the software project. TraCI4MATLAB was completed within this Sprint, with duration of approximately two months thanks to the best practices adopted from software re-engineering and related patterns. TraCI4MATLAB is described in detail in [1, p. 4]. After its introduction, TraCI4MATLAB has been improved in performance, and now features the latest SUMO commands, such as those related to pedestrian simulation. The successful implementation of TraCI4MATLAB is demonstrated in Sect. 4, where a pedestrian simulation is performed.

3.2   Sprint No. 2: Building the SUMOlib4MATLAB Library

SUMOlib tool is a *SUMO* library package wrote in Python that permits to understand *SUMO* networks; therefore, it was required to implement it in the *SUMO* and *MAT-LAB* integration process. The second Sprint focused in a re-engineering strategy to build *SUMOlib* for *MATLAB*, implementing a Graphic User Interface (GUI) for editing purposes. The re-engineering process found that *SUMOlib* is constructed using a SAX parser, which is not implemented in *MATLAB* up to now, then the conclusion was to build the component in *Java* and include it to the *Java* path in *MATLAB*. For generate traffic demands, based on object-oriented programming a component was coded to wrap the *SUMO DUAROUTER* application taking into account the use of turning probabilities. The developed component for demands uses as input a

MATLAB vector of vehicles demands and assigns it along an interval defined by the user, therefore permitting to define adjustable demands. Consequently, for *MATLAB* experimented researchers, the tool brings the mechanisms that simplify the definition of multimodal traffic demand for *SUMO* by using *SUMO*lib4MATLAB. The resulting main components can be obtained: the net component that is used to identify and extract the edges of interest in the *SUMO* network and the demand component that can be used to define different vehicle types and the corresponding traffic demand.

As in *TraCI4MATLAB* developed process, *SUMOlib4MATLAB* was developed by implementing a re-engineering process on *SUMOlib*. The *SUMOlib4MATLAB* design is illustrated in Fig. 1. Figure 1a shows the *UML* diagram describing how *SUMOlib4MATLAB* package depends on the implementation of *Java SAX*. Figure 1b, c illustrates the class diagrams (drawn in *UML* language) describing the net and sub-packages of demands, correspondingly. The figure only shows the most important attributes; nevertheless, *SUMO*lib4MATLAB supports all the attributes allowed by *SUMO*. The described flow class was conceived to support demand generation based in turning ratios at junctions via *JTRROUTER*, or defining origin and destination edges using *DUAROUTER*. More in detail, for demand based in turning ratios, it is required to create the turning probabilities using the `TurnProbability` class. In the same way, stops for public transport can be specified and multimodal traffic demand can be created by means of `Stop` and `VehicleType` classes, respectively.

3.3   Sprint No 3: The Extension of the SUMO Network Editor

One of the main objectives of MOYCOT project is to develop an interactive tool that allows editing and manipulating the different components of a traffic simulator in a friendly way. This implies the construction of a Graphic User Interface for *SUMOlib4MATLAB* that permits to view *SUMO* networks and modify objects interactively, showing drag-and-drop options and dialog boxes functionality. In order to build this solution, it is important to use the software pattern: Model–View–Controller (MVC) [5]. This technique splits the model of the data (Model) from the presentation (View) and provides an interface between the user (Controller) and the other two components.

The MVC pattern is broadly used and implemented in most of the current scalable Web and desktop applications. Numerous software libraries intended for the development of Graphic User Interface (i.e., GUI toolkits) embraced the MVC pattern using also enriched classes that significantly streamline the development process. Nevertheless, during the design and development process, we identify that *MATLAB* do not use a rooted MVC implementation and also was found that the implementation of enriched graphs that use drag-and-drop options is a complex task. Consequently, it

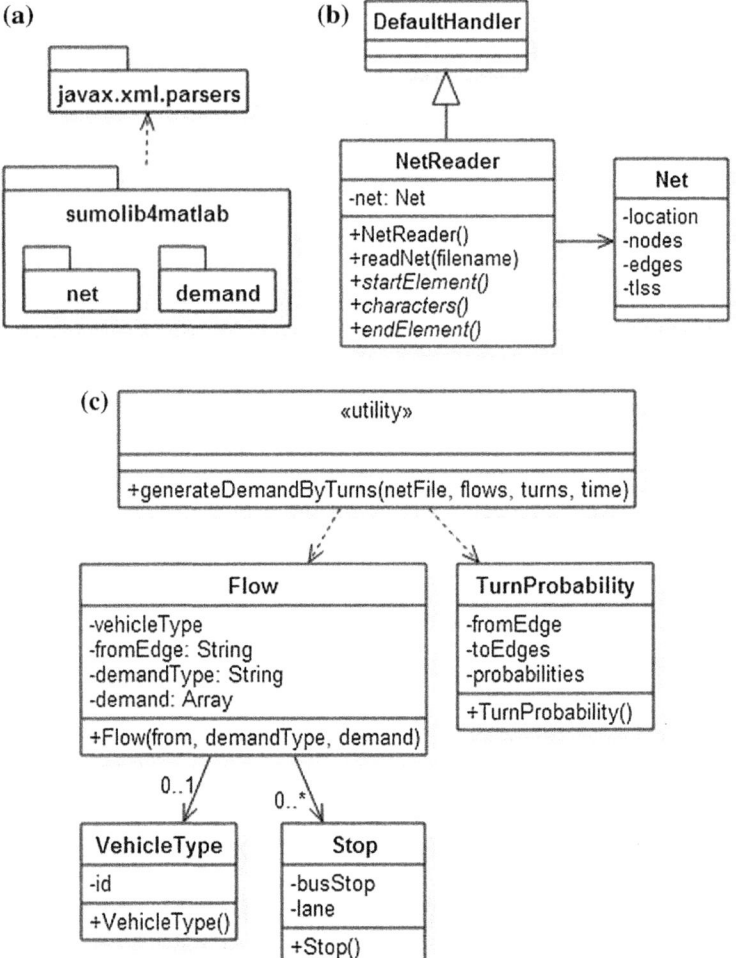

Fig. 1 Components of the *SUMO*lib4MATLAB: **a** The packages, **b** class diagram of net module, **c** class diagram of demand module

was required to explore alternative strategies to complete the requirements of Sprint No. 3. The NES—*Network Editor* for *SUMO* [3]—was identified as an option, being an open-source graphical Network Editor coded in *Qt/C++*, that was released in 2014.

The code development process in *Qt* has many advantages, even more in the GUI development, containing pre-defined models and views, and also a comprehensive event system constructed on "signals" and "slots."

The main features identified through a test process on NES and considered useful for this project are:

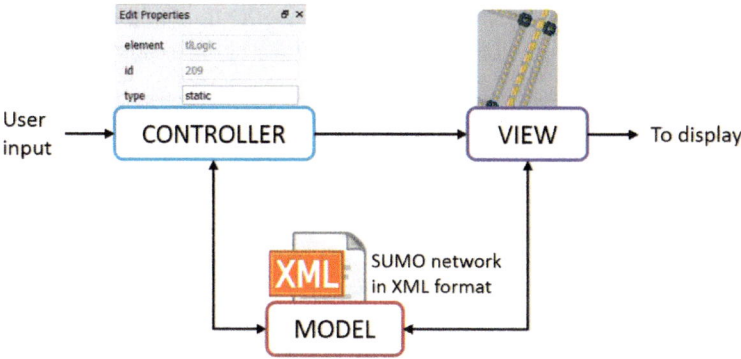

Fig. 2 Model–view–controller implementation of Network Editor for *SUMO*

- *SUMO* **networks visualization**. Figure 5b shows that NES displays the principal SUMO elements: edges, lanes, connections, and junctions in a unite graph view, preserving their geometry. Each element can be chosen so their options are displayed in "Properties View."
- **Editing options**. Graphical editing options include the alteration of junction, lane shapes, and edges. It is also possible to modify the element's options using the "Edit View."

Once more, the re-engineering strategy was implemented on NES looking for identify its design and to develop the new functionalities adequately. Particularly, it was found that NES uses the *SUMO* network file as a source, which is obtained by means of *NETCONVERT*, and produces a Document Object Model (DOM) instance corresponding to the model in the MVC pattern (Fig. 2). Furthermore, NES implements different views, including a main graph view, a standard Qt tree view that displays the objects hierarchy in the *SUMO* network and a couple of views that makes possible to visualize and edit a single *SUMO* object properties, specifically the "properties" view and the "edit" view, respectively. It is worth to clarify that some MVC implementations, like Qt, provide the view and controller functionalities in a single view class. Explaining why in Fig. 2 the "edit" view is in the controller's place.

The interpretation of the whole *SUMO* object structure is not an easy task especially when the *SUMO* network file is used as the model for implementing modules to fully create and edit *SUMO* objects. Moreover, *NETCONVERT* uses a heuristics set to be able to build geometries, connections, and right of way, to mention but a few, that would have to be taken into account. Moreover, the *SUMO* documentation advices the users to not edit the *SUMO* network file by hand, unless there is a deep understanding of its structure. This situation demanded more time that the originally assigned to the Sprint. Therefore, a direct way to accomplish the requirements of this

Sprint entails modifying the NES' model to include XML definitions [14] in place of the *SUMO* network file and to use *NETCONVERT* in the background to generate the final file.

The solution described is currently being implemented in MOYCOT project. Until now, two modules have been developed:

- **Network importing wizard**: This module is based in the importing capabilities of *NETCONVERT*. It provides a wizard that guides the user through the input, output, and processing option pages. The wizard is particularly useful when importing Open Street Maps (OSM). Figure 4 shows the network importing wizard.
- **Traffic lights' editor**: This module allows creating and editing traffic lights' program definitions, using a simple dialog box. In Fig. 6, the module is presented.

4   Results

The software tools described in the previous section were evaluated in the scenario described in Fig. 3a, showing the kilometer one—Via las Palmas: a main road in Medellín, connecting the city with its main airport: José María Córdova. In this road section, two traffic lights were installed during last year making possible the pedestrians crossing. Therefore, this scenario involves three transportation modes: public transport (buses and taxis), passenger vehicles, and pedestrians. The simulation involves recreating the pedestrian tutorial of *SUMO* [15], using TraCI4MATLAB. In this scenario, pedestrians can request to cross the street using a push button placed on the traffic lights.

(a)   **(b)**

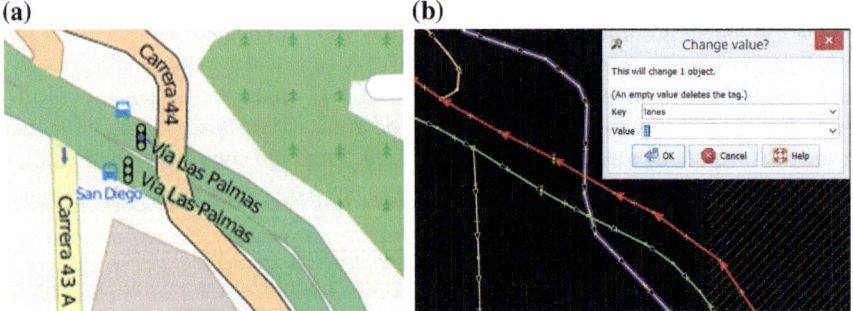

Fig. 3 First kilometer of Vía Las Palmas: **a** Open Street Maps' view, **b** JOSM editor view

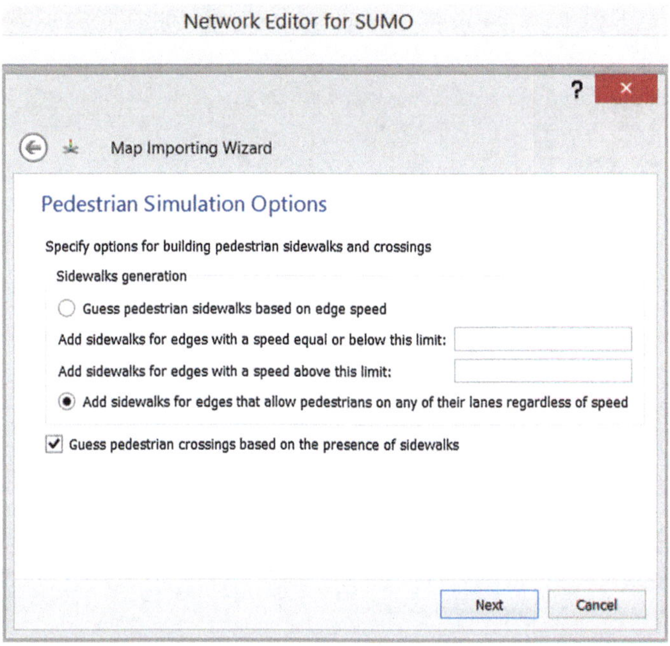

Fig. 4 Network Editor for *SUMO* (NES)—the importing wizard

For this scenario, using the Open Street Maps Web site, the area of interest was selected; then, we contrast to Google Street View in order to verify the numbers of lanes [16] using the JOSM editor, and this is described in Fig. 3b. Moreover, the traffic lights were detached in JOSM to test the traffic lights' editor.

It follows the use of the Importing Wizard of NES detailed in Sect. 3.3, to translate the OSM format to *SUMO* specification (.net.xml), as is showed in Fig. 4. The produced *SUMO* network is appreciated in Fig. 5. Figure 6 shows the creation of traffic lights of interest through the editor for traffic lights. The implemented traffic lights' functionality was then verified using SUMO *GUI*, as is displayed in Fig. 7.

The last step, the generation of traffic demand, for pedestrians, busses, and passenger vehicles was created for Las Palmas example, by means of the package for demands of *SUMOlib4MATLAB*. Afterwards, we proceeded to configure the simulation to evaluate *TraCI4MATLAB*. Figure 8 exhibits a screenshot of Las Palmas multimodal scenario. Mainly, the number of pedestrians on the walking areas beside Las Palmas and the green signal for vehicles in Vía Las Palmas westbound was obtained and plotted, as showed in Fig. 9, where the green signal takes a value of 1 whenever the traffic lights' phase representing the green signal for vehicles is active, and a value of 0 means that the traffic lights' state corresponds to any other phase.

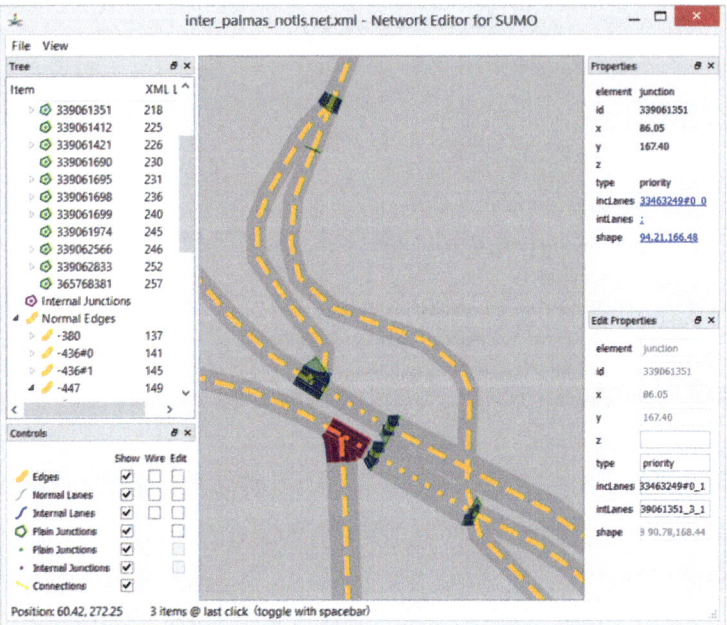

Fig. 5 *SUMO* network in Network Editor for *SUMO* (NES) obtained through the importing wizard

5 Conclusions

This chapter described the process to develop a set of tools that simplifies the construction of multimodal scenarios in *SUMO*. The software development process is implemented using the agile methodology SCRUM, and re-engineering available open-source packages. Following the methodology, the work required to complete the requirements defined by MOYCOT project was defined in three main Sprints, producing three fully functional software products: TraCI4MATLAB, *SUMO*lib4MATLAB, and a pair of components for the extension of Network Editor for *SUMO* (NES). These components make possible to import maps from the standard OSM and also editing programs of traffic lights. Besides the tools, this chapter provides a methodology to configure simulation scenarios for *SUMO* including the developed tools, and open solutions such as the JOSM editor and Open Street Maps, working in a complementary way. Finally, the full functionality of this solution was implemented in a reduced scenario in Medellín city.

The NES editor (Network Editor for *SUMO)* will be further developed to create and edit any *SUMO* object. Moreover, the demand generator graphical option will be developed in NES and extended to include the strategies used for demand generation in *SUMO* such as Traffic Assignment Zones (TAZ) and O/D matrices.

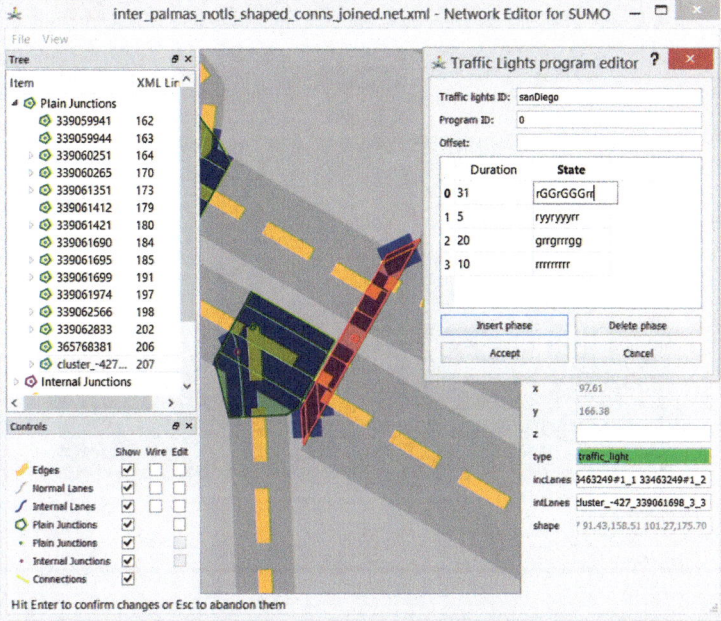

Fig. 6 Traffic lights' editor embedded on NES: implementing a traffic light program

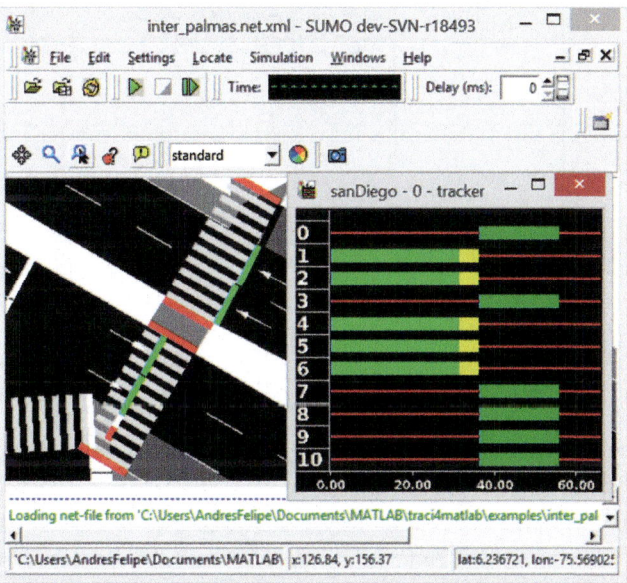

Fig. 7 Traffic lights' program obtained from the traffic lights' editor. (Now implemented for *NES)*

Fig. 8 Screenshot of Las Palmas multimodal scenario

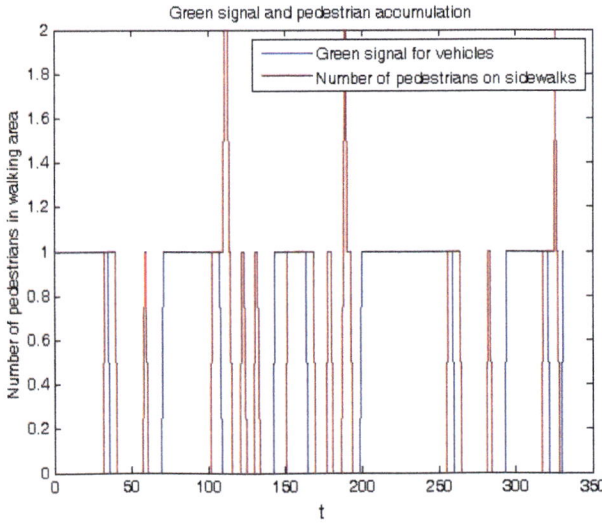

Fig. 9 Green signal and pedestrian accumulation on walking areas in Vía Las Palmas westbound

It is also planning to implement the software in a parallel architecture. The strategy is to deploy chained simulations with multiple instances. This will allow working in high concurrency and complex scenarios involving intricate traffic congestions that frequently affect cities with high density of vehicles.

Acknowledgements This work was supported by Proyecto Colciencias 111856934640 contrato–FP44842-202-2015: Modelamiento y Control de tráfico urbano en la ciudad de Medellín Fase 2. Convocatoria 669. Código Hermes: 25374.

References

1. Acosta A, Espinosa J, Espinosa JE (2014) TraCI4Matlab: re-engineering the python implementation of the TraCI interface. In Proceedings of the 2nd SUMO user conference SUMO2014, Deutsches Zentrum für Luft - und Raumfahrt e.V, pp 145–156
2. Acosta A, Espinosa JE, Espinosa J (2015) Developing tools for building simulation scenarios for SUMO based on the SCRUM methodology In Proceedings of the 3rd SUMO user conference SUMO2015, Deutsches Zentrum für Luft - und Raumfahrt e.V, pp 23–35
3. editor4sumo (2014) *SourceForge*. http://sourceforge.net/projects/editor4sumo/. Accessed: 28 Aug 2014
4. Qt Cross-platform application & UI development framework. http://www.qt.io/. Accessed: 22 Jun 2015
5. Krasner GE, Pope ST et al (1988). A description of the model-view-controller user interface paradigm in the smalltalk-80 system. J. Object Oriented Prog 1(3), 26–49
6. MOYCOT, MOYCOT (2015) http://gaunal.unalmed.edu.co/moycot/main/ Accessed 28 Jan 2015
7. Jalali S, Wohlin C (2010) Agile practices in global software engineering-a systematic map. In 2010 5th IEEE international conference on global software engineering (ICGSE). pp 45–54
8. Dyba T, Dingsoyr T (2009) What do we know about Agile software development? IEEE Softw 26(5):6–9
9. Chikofsky EJ, Cross IJH (1990) Reverse engineering and design recovery: a taxonomy. Softw IEEE 7(1), 13–17
10. Demeyer S, Ducasse S, Nierstrasz O (2002) Object-oriented reengineering patterns. Morgan Kaufmann, San Francisco
11. Schwaber K, Sutherland J (2011) The scrum guide Scrum Alliance
12. Seminario Internacional DE TRÁFICO Y TRANSPORTE. https://sites.google.com/site/moycot1/home. Accessed 03 Feb 2015
13. TraCI - Sumo. http://sumo.dlr.de/wiki/TraCI. Accessed: 01 Apr 2015
14. Networks/Building Networks from own XML-descriptions - Sumo. http://sumo.dlr.de/wiki/Networks/Building_Networks_from_own_XML-descriptions. Accessed: 24 Jun 2015
15. Tutorials/TraCIPedCrossing Sumo. http://sumo.dlr.de/wiki/Tutorials/TraCIPedCrossing. Accessed: 22 Jun 2015
16. Street View - Google Maps. https://www.google.com/maps/views/u/0/streetview?gl=us. Accessed: 22 Jun 2015

Multi-resolution Traffic Simulation for Large-Scale High-Fidelity Evaluation of VANET Applications

Manuel Schiller, Marius Dupius, Daniel Krajzewicz, Andreas Kern and Alois Knoll

Abstract This paper presents an approach for coupling traffic simulators of different resolutions in order to conduct virtual evaluations of advanced driver assistance systems based on vehicular ad hoc networks that are both large scale and a high fidelity. The emphasis is put on the need for such an attempt to satisfy the constraint of performing simulations in real time. Both the methods to accomplish this as well as the resulting performance are described.

Keywords V2V communication · ADAS · Multi-resolution simulation

1 Introduction

Vehicular ad hoc networks (VANETs) have attracted a lot of research attention over the last years due to having the potential for improving traffic safety, efficiency, and driver comfort. A high variety of applications, commonly referred to as advanced driver assistance systems (ADAS), such as cooperative driving and subsequently automated driving, can only be enabled through wireless communication between the vehicles on the road.

Before deployment of such systems, which often exhibit safety-critical features, in series production on a large scale, a lot of effort has to be put into testing and

M. Schiller (✉) · A. Knoll
Lehrstuhl Für Echtzeitsysteme Und Robotik, Technische Universität München,
Boltzmannstr. 3, 85748 Garching, Germany
e-mail: manuel.schiller@in.tum.de

M. Dupius
VIRES Simulationstechnologie GmbH, Grassingerstraße 8, 83043 Bad Aibling, Germany

D. Krajzewicz
Institute of Transportation Systems, German Aerospace Center, Lilienthalplatz 7, 38108
Braunschweig, Germany

A. Kern
AUDI AG, Ettinger Straße, 85057 Ingolstadt, Germany

© Springer International Publishing AG, part of Springer Nature 2019
M. Behrisch and M. Weber (eds.), *Simulating Urban Traffic Scenarios*,
Lecture Notes in Mobility, https://doi.org/10.1007/978-3-319-33616-9_2

validation. Although real test drives using physical test beds of prototype vehicles offer the highest degree of realism, the large amount of financial, material, and human resources needed to perform large-scale and extensive testing of vehicular networks renders their use rather impossible. Due to this, simulations are employed for obtaining a view of the performance of such solutions in large-scale virtual environments. As well, simulation-based evaluation techniques particularly allow testing those complex systems in a wide variety of dangerous and critical scenarios without putting humans and material at risk and at low costs.

In the automotive industry, the use of simulation is well established in the development process of traditional driver assistance and active safety systems. However, the current emphasis is primarily on the simulation of individual vehicles at a very high level of detail [1]. When investigating and evaluating the performance of ADAS based on vehicular communication, this isolated view of a single vehicle alone or a small number of vehicles in the simulation is not sufficient anymore. Potentially, every vehicle equipped with wireless communication technology is coupled in a feedback loop with the other road users participating in the vehicular network, and therefore, the number of influencers that need to be taken into account is drastically increased.

These considerations lead to a trade-off between the accuracy in terms of the simulated level of detail of each vehicle and the scalability in terms of the number of vehicles that can be simulated with the available computing resources. In this paper, we present an approach to solve this trade-off by coupling multiple resolutions of traffic simulations to get highly accurate simulation results where it is necessary and simultaneously achieving an efficient simulation of large-scale scenarios. Such combinations pose several technical problems which were already reported in the literature; see [2] or [3]; e.g., mostly, they are due to difference in how vehicles are represented or allocated in the involved simulators. Some solutions to similar technical issues could be developed and will be discussed in the following.

The rest of this paper is organized as follows. The testing and evaluation of real-world implementations of ADAS impose a certain set of additional requirements, which are discussed in Sect. 2 before giving an overview of the related work. Section 3 describes the concept of our multi-resolution traffic simulation approach. Some specific issues found during the implementation are discussed in Sect. 4, including their solutions. In Sect. 5, we evaluate the performance by means of an exemplary scenario. In Sect. 6, we discuss the limitations of the approach and conclude the paper in Sect. 7.

2 Background and Related Work

Before we proceed to the discussion of related work, it is essential to illustrate our scope and area of application. In order to evaluate and validate real implementations of ADAS in a simulated, virtual environment, both state and behavior of the vehicles must be modeled and simulated in high fidelity. We define the term *high fidelity* as a three-dimensional problem.

To substitute the real vehicle by its simulated counterpart, the virtual vehicle must provide its state variables in a *sufficient range*, in a *sufficient precision*, and in a *sufficient temporal resolution*. The concrete manifestations of these three requirements depend on the respective use case. For example, a cooperative adaptive cruise control (CACC) system will need, among others, the dynamic state of the vehicle (e.g., speed, acceleration, position), power train state, RADAR sensor values as well as the input from the wireless communication channel at different sampling rates [4]. If any of these requirements are not met, e.g., a necessary state variable is not covered by the simulation model, simulative testing cannot be performed.

Since the range of wireless communication technology is higher than what can be achieved through conventional sensors, even vehicles which are farther away can act as relevant information sinks and sources and therefore influence the driving assistance system as well as the road traffic system as a whole. In order to capture these effects in the simulation, a naive approach would be to simply scale existing, high-detailed simulations by increasing the amount of vehicles in the simulated area. However, these high-detailed simulations are extremely computationally intensive and hence are not suitable to perform evaluations of large-scale scenarios in a reasonable amount of time. This also prohibits their use in hardware-in-the-loop simulations where a real-time constraint must be fulfilled [1].

This additional timing requirement conflicts with the aforementioned three dimensions of simulative high fidelity. There are numerous references that deal with similar issues in the fields of traffic simulation, VANET simulation, and multi-resolution simulation. In the following, we give a brief overview of those research areas in order to state the fundamentals of this investigation.

2.1 Traffic Simulation

Road traffic simulations can generally be subdivided into the following four categories according to the level of detail [5, 6]: macroscopic, mesoscopic, microscopic, and nanoscopic. While macroscopic flow models describe traffic at a high level as the aggregate traffic flow, microscopic simulations model the behavior and interactions of each simulated entity individually with specific state variables such as position, speed, and acceleration. Mesoscopic models are medium-detailed models where traffic is usually represented by queues of vehicles. In nanoscopic models, which are also referred to as submicroscopic models, an even higher level of detail is achieved through the subdivision of each vehicle in multiple subunits. This allows to model, for example the vehicle dynamics, complex decision processes of the driver or the interaction with the vehicle surroundings more accurately. The necessary amount of computation time for the traffic simulation rises considerably with the increasing degree of detail.

2.2   VANET Simulation

VANET Simulation: The usual strategy to simulate VANETs found in the literature is to bidirectionally couple a network simulator and a microscopic traffic simulation. Following this approach, the interactions between road traffic and network protocols are represented and the mutual impact can be explored [7, 8]. The majority of research publications focuses on the investigation of low-level networking subjects such as medium access [9] or rather high-level concepts of applications such as reducing CO_2 emissions [10]. For this kind of studies, it is sufficient to apply realistic mobility patterns originating from the microscopic traffic simulation. In this bird's eye view of the overall system, it is not necessary to model individual cars in the high level of detail mentioned above because the evaluated systems change each vehicle's behavior at a timescale of seconds to hours. This is but replicated by microscopic simulations at a sufficient quality level. Therefore, the three requirements of high fidelity in terms of modeling and simulation of individual vehicles need not be considered when simulating VANETs on such an abstract level.

2.3   Multi-resolution Simulation

Multi-resolution modeling (MRM) is defined as the combination of different models of the same phenomenon at different levels of resolution which are then executed together [11]. This methodology allows to find a good balance between simulation accuracy and computing resources. High-resolution models, which provide accurate simulation results at the cost of high computational efforts, are only applied in limited areas of interest whereas the major part of the simulation is handled by less accurate but also less resource consuming low-resolution models. However, not only the difference in execution speed can be exploited but also the fact that low-resolution models tend to give a better overall understanding of the system under examination because of their rather abstract view of the big picture.

  MRM has been successfully applied in road traffic simulation. In [12], a combination of a microscopic simulation modeling, the inter-vehicle interactions, and a computationally less expensive macroscopic simulation applied to freeways is described. This allows a scalable, yet accurate investigation of traffic flow in large-scale networks. This approach is not applicable for VANET simulations since due to the flow-based simulation in macroscopic models accurate vehicle positions are missing which is crucial when simulating vehicular networks. In [13], a coupling of two different microscopic traffic simulators of different accuracy is implemented for VANET simulation to exploit the difference in execution speed. In [13] and in [14], the areas of interest are fixed throughout the simulation; for example, road intersections are simulated at a higher level of detail than freeways. In contrast to that, the areas of interest are not statically fixed in [15] but rather depending on the simulation context.

2.4 Combining Traffic and Driving Simulation

A rather recent research direction is the combined simulation of both traffic and human-in-the-loop driving simulation. A co-simulation of a microscopic traffic simulation and a driving simulator [16] respectively a robotics simulator [17] has been investigated. In these approaches, the behavior of a fixed subset of all vehicles is remote controlled through an external simulator, which allows to have a predefined number of detailedly simulated vehicles to be surrounded by a large number of microscopically simulated vehicles.

3 Developed Multi-resolution Traffic Simulation

In the following, the realized system is presented. The involved simulations are presented, first. Then, the technical realization of a co-joint traffic scenario execution is given.

3.1 Utilized Simulators

In the following, we describe the concrete manifestations of the simulators which we combine using the above-described concept to achieve a multi-resolution traffic simulation.

3.1.1 Microscopic Traffic Simulator—SUMO

We chose to use Simulation of Urban MObility (SUMO) [18] as the traffic simulator responsible for the simulation of the LRA. SUMO is a microscopic, space-continuous, and time-discrete simulator. While it is employed in a wide range of research domains, its most notable use is shown in a high number of research papers regarding VANET simulations [19]. SUMO is well known for its high execution speed as well as for its extensibility. Due to its efficiency, which is partly achieved through its simplified driver model [20], SUMO is ideally suited to simulate a high number of vehicles residing in the LRA.

3.1.2 Nanoscopic Traffic and Vehicle Simulator—VIRES Virtual Test Drive

We employ the nanoscopic traffic and vehicle simulator VTD for the simulation of the high-resolution vehicles. VIRES Virtual Test Drive (VTD) has been developed for the

Fig. 1 3D visualization of a
simulated RADAR sensor in
VTD

automotive industry as a virtual test environment used for the development of ADAS
[21]. Its focus lies on interactive high-realism simulation of driver behavior, vehicle
dynamics, and sensors. VTD is highly modular, so any standard component may be
exchanged by a custom and potentially more detailed implementation. Its standard
driver model is based on the intelligent driver model [22]; however, an external
driver model may be applied if necessary. The same concept applies to the vehicle
dynamics simulation, where the standard single-track model can be substituted by
an arbitrarily complex vehicle dynamics model adapted for specific vehicles. Each
simulated vehicle can be equipped with arbitrary simulated sensors, for example a
RADAR sensor, which is shown in Fig. 1 .

   While VTD is designed for online operation, it is, however, not suited to simulate
a large number (i.e., thousands) of vehicles with respect to real time due to the details
and complexity of the simulation. Therefore, only the EGO car as well as the vehicles
residing in the HRA are simulated by VTD.

3.2 Coupling Concept

3.2.1 Offline Preprocessing

Both simulators rely on different data formats representing the modeled road network.
In order to be able to run a co-simulation of both simulators, the underlying data basis
has to match. VTD uses the OpenDRIVE format to specify the road network. This
specification models the road geometry as realistically as possible by using analytical
definitions. SUMO on the other hand approximates the road network geometry by
line segments. There are additional differences in the modeling of intersections and
lane geometries. To achieve a matching database, we convert the road network in an
offline preprocessing step from OpenDRIVE to the file format SUMO supports.

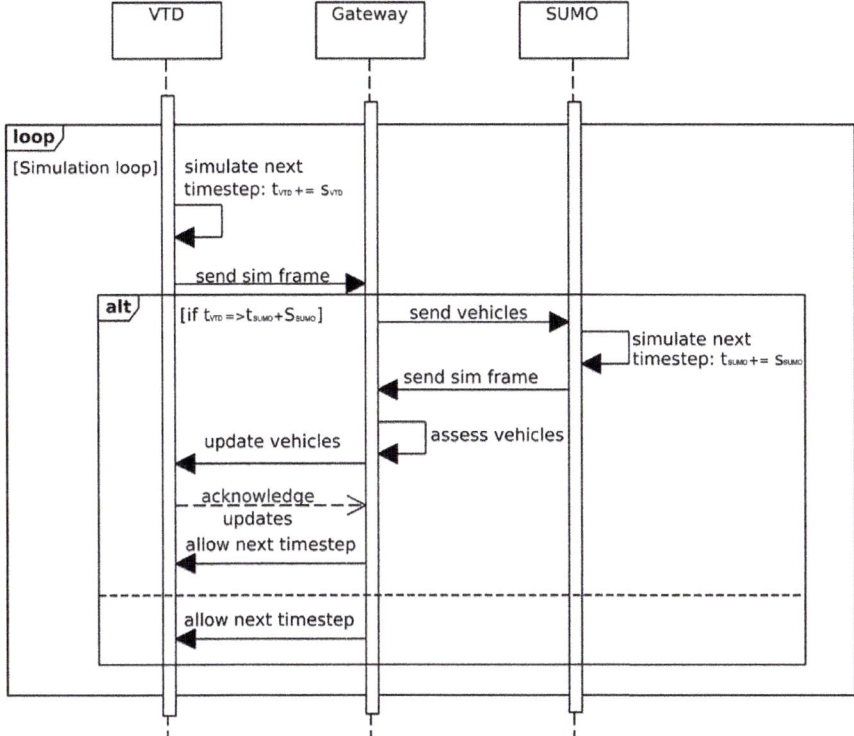

Fig. 2 Synchronization

Fig. 3 Dynamic partitioning
of simulated area

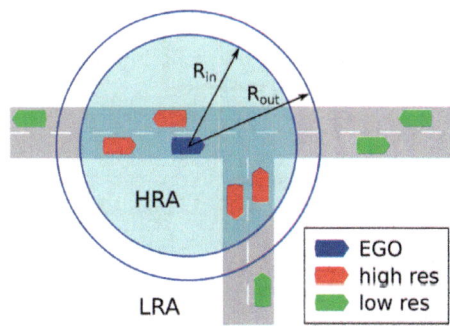

3.2.2 Online Coupling and Synchronization

The coupling of the simulators at simulation runtime is based on the master–slave principle. Figure 5 shows this sequence of operations during a single simulation step, in which VTD and SUMO can operate with different temporal resolutions. s_{VTD} is the length of a timestep for the HRA, whereas s_{SUMO} is the length of a timestep for

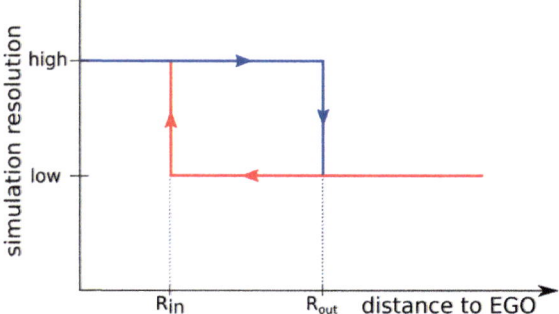

Fig. 4 Hysteresis control of the simulation resolution

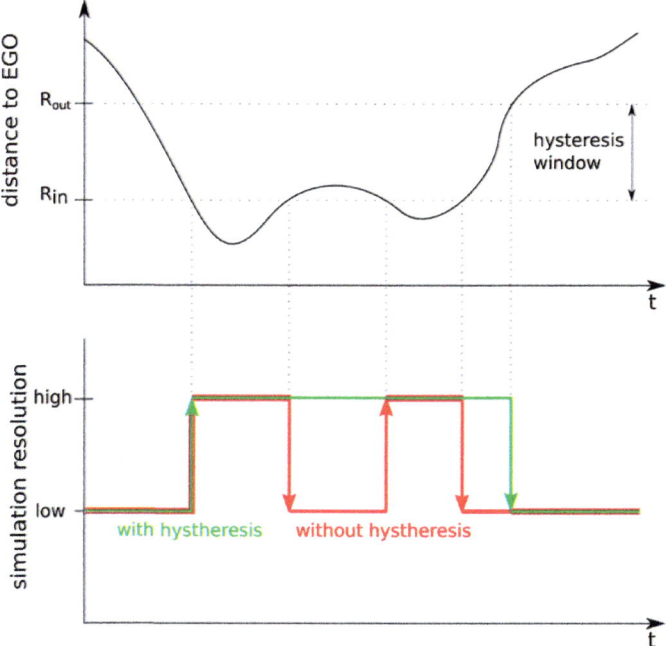

Fig. 5 Comparison of simulation resolution switching

the LRA. Typically, the nanoscopic simulation is run at a higher frequency than the microscopic one. t_{VTD} respectively t_{SUMO} denote the local simulation time in each simulator. At the beginning of each simulation step, a new timestep is simulated in VTD. If the next timestep has been reached for SUMO and therefore the condition $t_{VTD} \geq t_{SUMO} + s_{SUMO}$ is fulfilled, the state of the high-resolution vehicles is sent to SUMO through a gateway. It then triggers the simulation of the next timestep in the low-resolution model, and as a result, the positions of the low-resolution vehicles are passed back. These vehicles are now classified according to Sect. 4.1, and, if applica-

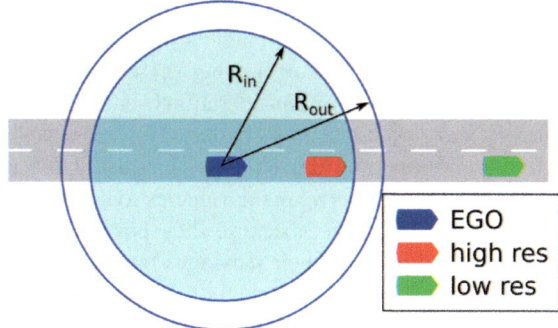

Fig. 6 Traffic flow consistency issue at the border of HRA

ble, the change of resolution is performed for individual vehicles. When an exchange of a vehicle between the simulators happens, the previously mentioned inherent difference in the underlying road network may cause problems if a vehicle cannot be mapped based on its position on a specific lane due to difference in accuracy. This is especially true for complex intersections which are modeled quite differently.

After all resolution changes have successfully been completed, the simulation is unblocked again and the next timestep can be simulated. This synchronization is very important to ensure reproducible simulation results across multiple simulation runs.

3.3 Implementation

The gateway depicted in Fig. 2 is implemented as a dynamic plug-in for VTD written in C++. It uses the TraCI network interface [23] provided by SUMO to control the vehicles of the microscopic simulation.

4 Realized Solutions

4.1 Dynamic Spatial Partitioning of the Simulated Area

Our approach aims to couple traffic simulation models of different resolutions at dynamic regions of interest. Contrary to conventional traffic simulation, we are not interested in investigating a large number of vehicles from a bird's perspective, but the focus is rather on a single vehicle or a limited number of vehicles which are used to conduct a test drive in a virtual environment. In the following, we will refer to this kind of vehicle as the EGO car. The ADAS under investigation is imagined to be on board of such an EGO car. The simulated measurements and sensor values are fed into the ADAS. Depending on its type and its use case, the respective ADAS directly or indirectly influences the vehicle's state and behavior.

As stated before, all surrounding vehicles both near and far away from the EGO car need to be taken into consideration because of the wide range of transmission on the wireless communication channel. However, we can distinguish between highly relevant and less relevant vehicles. This distinction is based on the criterion of the respective distance between the vehicles to the EGO car. Nearby vehicles are inherently of more relevance because they pose a higher danger in terms of possible collisions and because their messages transmitted on the vehicular network are of higher importance due to the vicinity of their origin.

Based on this distance criterion, an area of interest is defined which is centered around the EGO car and in which the defined simulative high-fidelity requirements must be fulfilled. Since the EGO car is driving continuously through the virtual environment, this area of interest is being moved along likewise. We therefore partition the global area of the simulation dynamically into a high-resolution area (HRA) and a low-resolution area (LRA). Figure 3 shows a schematic view of the dynamic spatial partitioning. There the HRA is defined as the area of a circle which is centered around the EGO vehicle. Red vehicles are within that circle and are therefore simulated in high resolution by the involved nanoscopic simulator, whereas the green vehicles are outside of the circle and are consequently simulated in low resolution by a microscopic simulation. All vehicles exist in the microscopic simulation, but in the nanoscopic simulation only the high-resolution vehicles are contained and their movements are applied to their proxy counterparts in the microscopic simulator.

Due to the dynamic nature of road traffic, the EGO car, the high-resolution vehicles as well as the low-resolution vehicles are permitted to move continuously. The classification of the assigned resolution mode is therefore performed after each timestep of the simulation. Vehicles for which the classification has led to a change in resolution are transferred to the appropriate simulator. This change of resolution is possible in both directions at every timestep. However, since the HRA is defined to be centered around the EGO car, it is always simulated in high resolution.

In order to prevent vehicles which are close to the boundary between HRA and LRA from oscillating very frequently between the two resolution areas, a hysteresis controller as depicted in Fig. 4 is applied in the classification process. As shown in Fig. 3, the two thresholds R_{in} and R_{out} are defined. A vehicle is transferred into the high-resolution simulation only if its distance to the EGO car falls below the value of R_{in}. The exchange back to the low-resolution simulation is carried not out until the threshold R_{out} is exceeded.

The difference is shown in Fig. 5 by an exemplary trajectory. Without the hysteresis, the change in resolution is performed multiple times, whereas when applying the hysteresis controller the vehicle stays in the high-resolution model while being close to the boundary.

The extent of R_{in} defines the circumcircle in which vehicles are simulated by the nanoscopic simulator. This value needs to be determined separately for each application scenario; for example, in an urban scenario due to the expected low traffic speed, a lower value can be used than what is suitable for a freeway scenario.

As an alternative to the definition of a fixed size, the value of R_{in} can be determined dynamically at simulation runtime, for example, based on traffic speed, volume of traffic, or even depending on the currently available computing capacities.

The gap $R_{out} - R_{in}$ defines the size of the hysteresis window for the dynamic change between the two simulators. R_{out} can be selected in both absolute and relative terms in relation to R_{in} and exhibits a lower dependence with regard to specific scenarios and traffic speeds.

Our approach of dynamic spatial partitioning of the simulated area enables us to allocate the processing resources between the simulators. As the focus is only on a relatively small region of the simulated area, we achieve both a simulative high fidelity in this region of interest and the simulation of large-scale scenarios.

4.2 Generalization of the Approach

While the description in Sect. 4.1 focuses on the simulation of a single EGO car, the concept can be generalized as follows. Generally speaking, multiple EGO cars can be simulated analogously and can exist either in separate or in joined areas of high resolution. The dynamic spatial partitioning concept is not necessarily limited to having only two different simulators forming the multi-resolution simulation, but even more simulators could be added to the synchronization scheme in Fig. 4.

The definition of the area of interest is also not limited to a circle. The circle was chosen due to its simple definition and fast distance calculations in the classification process, but the HRA could be represented by arbitrary complex shapes. As another generalization, the classification process does not need to be based on pure geometrical calculations only but could also be enriched with logical conditions. For example, on a freeway vehicle driving on oncoming lanes could be excluded due to their limited relevance to the EGO car.

4.3 Exchange of Vehicle Positions and States

One of the most problematic subtasks in achieving the coupling was the correct allocation of vehicles within the SUMO road network, given the information available in VTD. Within SUMO, every vehicle is assigned to a specific lane and has a certain position along it. Within VTD, vehicles can move "freely": Overtake on the opposite lane, leave a lane when turning, etc. Thereby, a direct exchange of position information is not possible:

- **VTD determines the lane and sends it to SUMO**: Because vehicles are not assigned to lanes within VTD, VTD can only determine the lane below the vehicle. In case of overtaking at the opposite lane or within intersections where paths crossing the intersection (modeled as "internal" lanes in both models) may overlap, this information is counterproductive.

- **SUMO maps vehicles based on positions sent by VTD**: It is assumed that the differences in processing road networks (see above) yield an incorrect mapping. Mapping the position on the opposite lane is likely to happen too often. As well, the ambiguity of intersection-internal lanes remains.

As a consequence, different heuristics to determine the correct lane the vehicle shall be located at in SUMO were implemented and tested. At first, additional information tried to be generated by building a mapping from OpenDRIVE lane IDs to the according ones used in SUMO. This has been implemented within the network importing tool. Though, this neither reduces the ambiguity at intersections nor is it intuitive enough for users—the mapping was often not found in customer networks, probably because it has to be enabled using an additional option.

Another approach tried to find the position following the fixed route information given in SUMO. For dealing with vehicles that are controlled by a human being and leave their route, a threshold was used. When the given VTD position differs from the SUMO position found along the route by more than this threshold, a rerouting is assumed. However, employing this method resulted in halting vehicles which were waiting for the threshold to be reached for being moved at their current edge that is not a part of their route.

For this reason, the currently implemented method treats the given information as hints. Besides the position and heading angle sent by VTD, the conformity with the route and continuation of the used edge are taken into regard. All this information is individually weighted for obtaining a conformity measure for each edge geometry part that is within the investigated area around the sent VTD position. The best matching lane position is eventually chosen. This approach is the most robust one that was investigated so far.

Within SUMO, vehicles controlled by VTD are marked as such. They are ignored within the simulation loop, first, and their positions are updated using the submitted ones at the end of a simulation step.

It was assumed at the beginning of the project that a correct mapping of positions is crucial and the position difference should therefore be kept as small as possible. But tests have shown that a too big detail grade highly reduces the simulation speed. An online execution of both simulators could not be guaranteed. The reason is the need to find the according position along the lane. As stated, this can only be accomplished by computing the minimal geometrical distance to all lines of a SUMO lane's geometry. Therefore, an option for defining the accuracy of the road network was added to the network importer, which handles the conversion from the OpenDRIVE format to the SUMO network format.

Similar issues had to be tangled on the VTD side. The OpenDRIVE format allows to define lanes of a variable width—e.g., growing along their length at on-ramps. SUMO does not support variable lane widths. It may thereby happen that a vehicle is placed at a lane position that is too narrow. Of course, this may happen due to the geometrical differences as well. Such off-road cases are caught and handled by VTD.

4.4 Maintaining Traffic Flow Consistency

As previously stated, the dynamic spatial partitioning is motivated by performance improvements through the reduction of highly detailed vehicles which are located in proximity to the EGO car. This results in only a limited subset of the overall simulated vehicles to be represented by the high-detailed models. Since only this subset exists within VTD at a given time, the surrounding traffic flow simulated by the low-resolution model is not known to the vehicles inside the HRA. This missing information can result in consistency issues which are described using the exemplary scenario depicted in Fig. 6.

The scenario consists of three vehicles: The EGO car, a high-resolution vehicle as well as a low-resolution vehicle are driving along a straight road. The high-resolution vehicle which is located near the boundary of the HRA does not have a preceding vehicle. The underlying driver model therefore assumes a free road ahead as the low-resolution vehicle is not visible to it. This results in a higher driving speed as if the vehicle in front was known to the driver model. There are two possible outcomes that depend on the particular relative positioning of the three vehicles:

- The high-resolution vehicle "flees" out of the HRA and is then transferred to the low-resolution simulator, and the preceding vehicle becomes abruptly visible to it.
- Since the EGO car follows the high-resolution vehicle at a high speed, the HRA which is centered around the EGO car moves along and captures the low-resolution car which is then transferred to the high-resolution simulator. This also results in an abrupt change in visibility to the high-resolution car.

These sudden changes in visibility cause the driver model to decelerate sharply and cause artificial shock waves which are only caused by the missing information of the surrounding traffic flow. Figure 7 shows the trajectories of the three cars for the second case where the slower low-resolution car is caught up and is transferred to the high-resolution model at around $t = 47$ s. The acceleration of the high-resolution car shows a sharp discontinuity to accommodate the fact that the preceding vehicle suddenly materializes.

An efficient solution to this problem takes advantage of the fact that car-following models are typically governed by collision-free motion of the vehicles. This absence of collisions is guaranteed by bounding the velocity of each vehicle by a safe velocity [25]. For all high-resolution vehicles which do not have another preceding vehicle

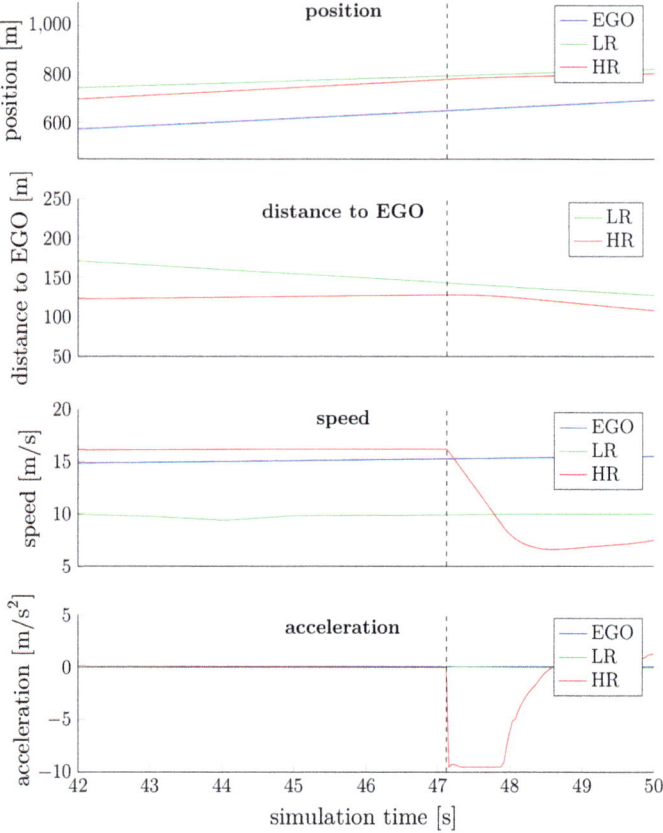

Fig. 7 Change of simulation resolution of the low-resolution model causes abrupt deceleration of the HR car

inside the HRA in front of them, the next vehicle outside of the HRA is searched. If one is found, the safe velocity can be calculated based on the distance and the current velocity. This calculation can happen either within SUMO or within VTD. As the time step for SUMO is chosen to be rather coarse, the upper bound values would have to be interpolated. Since the underlying formula can be efficiently implemented and evaluated within VTD, this solution is preferred. Figure 8 shows the trajectories for the same scenario with the velocity bound applied. The high-resolution vehicle now starts decelerating before the low-resolution vehicle is within the HRA, which results in smooth speed and acceleration profiles without producing artificial shock waves.

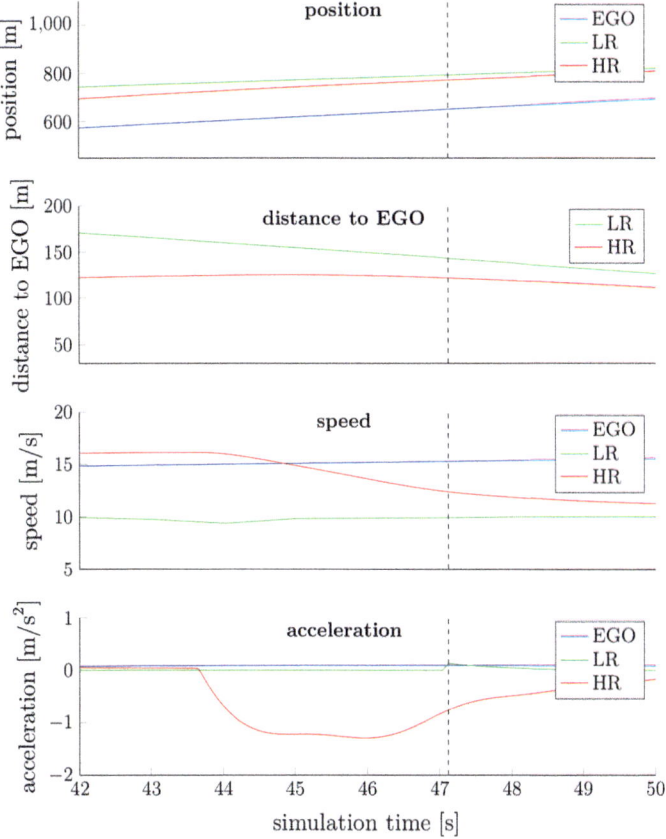

Fig. 8 Applying the upper bound velocity for the HR car resolves the consistency problem

5 Evaluation

5.1 Scenario and Simulation Setup

A synthetic scenario was created for testing the coupling concept and evaluating its performance. It consists of a single straight road running west to east with a length of 50 km and two lanes, one for each direction. Each lane is configured to have a constant inlet of 1000 vehicles per hour heading either east or west. The EGO car is located near the start of the road. It drives from west to east and is followed by a traffic flow and heading to the oncoming traffic flow. This artificial road was first modeled in the OpenDRIVE format and was then converted to the SUMO road network format.

We performed two series of experiments. In the first series, the nanoscopic traffic simulator VTD was applied to the whole simulated area. In the second series, we used the described multi-resolution concept to partition the simulation area between VTD and SUMO. We chose a timestep of $s_{VTD} = 20$ ms for the high-resolution area in VTD and a timestep of $s_{SUMO} = 1$ s for the low-resolution area in SUMO. The hysteresis thresholds which define the dynamic area of interest were set to $R_{in} = 500$ m and $R_{out} = 550$ m.

Both simulators are executed on the same computer which is equipped with an Intel Xeon CPU E5-1620 at 3.60 GHz and 16 GB of RAM. The operating system is Ubuntu Linux 14.04 with a 3.13 64-bit kernel. We used VTD version 1.4.1 with 3D rendering disabled and a custom branch of SUMO v0.21 that incorporated an in-work version of the methods needed to exchange vehicles between VTD and SUMO.

5.2 Performance Evaluation

We measured the duration it takes to perform each simulation step over the simulation period of the 1800 s, while the number of vehicles is constantly being increased. Each series consists of five separate simulation runs to account for fluctuations in the measured execution times. To illustrate the trends of the measurements more clearly, the moving average is also displayed in the following figures.

Figure 9 shows the performance development of the nanoscopic simulation while increasing the simulated vehicle count over the simulation period. The duration of each simulation step is almost constant up to a count of 70 vehicles. Until then, the duration is around 12 ms, which is less than the timestep length of 20 ms and therefore yet fulfills the real-time constraint. At around 150 vehicles, the duration is beyond these 20 ms and real-time simulation is not possible anymore. With increasing vehicle count, the duration for each timestep also considerably increases and reaches 180 ms at the end of the simulation period. This results in an increase of factor 15 compared to the amount of computation time at the beginning of the simulation. The overall simulation took over 120 min to complete, which is four times more than the simulated time.

Figure 10 shows the performance development of the multi-resolution simulation in the same simulation scenario over the same simulation period. While the total vehicle count is increased the same way as in the pure nanoscopic simulation, the separately plotted nanoscopic vehicle count illustrates the amount of cars which are within the high-resolution area. It shows that reducing the nanoscopic model's area of interest fulfills the aim of reducing the overall simulation time. After a local maximum of 11 nanoscopic cars is reached, this count decreases slowly since slower vehicles are left behind the faster moving EGO car. At around simulation time 1350 s, the two traffic flows from each end of the road meet in the middle of the road, which then increases the nanoscopic vehicle count. However, due to the limited extent of the HRA the nanoscopic vehicle count does not exceed a certain limit, which for the given configuration is at around 27 vehicles. The duration for the timesteps stays on average constant around 12 ms, so it can be seen that the overhead resulting from the coupling of the two simulators is negligible as is the execution time of the

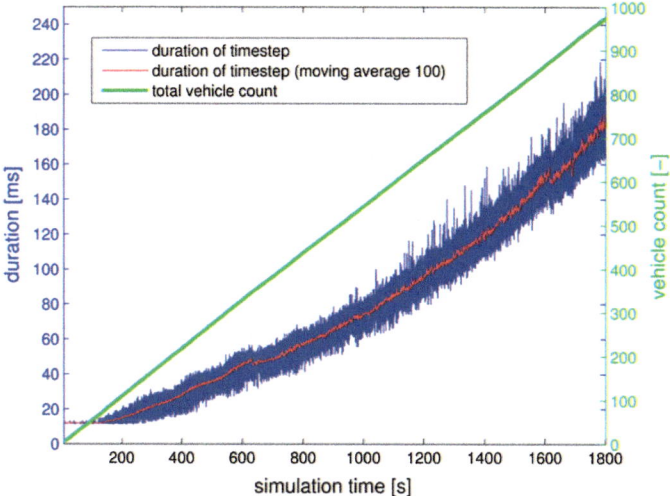

Fig. 9 Simulation performance—nanoscopic simulation only

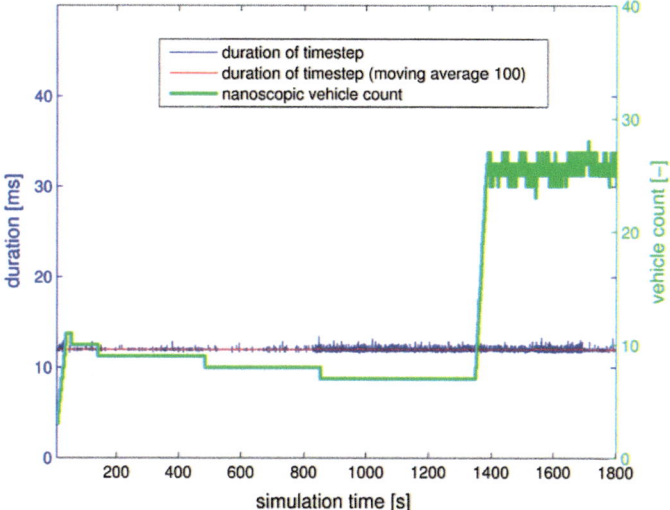

Fig. 10 Simulation performance—multi-resolution simulation

microscopic simulator due to its less detailed yet much more efficient simulation model. The overall simulation took less than 18 min to complete, so the simulation is faster than real time by factor 1.66. The real-time constraint is fulfilled throughout the whole simulation period.

6 Discussion

The above-shown performance evaluations state that the multi-resolution approach leads to the desired goal of a highly detailed simulation in the region of interest while maintaining a high overall performance. However, there are additional concerns, which we cover in the following.

When switching between resolutions and simulators, maintaining simulation consistency is of utmost importance [24]. The change of resolution from high to low can generally be handled rather simply by losing information through a transformation function. The opposite direction though is problematic. The presented approach will face a consistency problem when using vehicle dynamics models with a higher level of detail. When switching from low to high resolution, only a minimal subset of the state variables (position and speed) is available. The remaining state variables (pitch and roll angle of the car body, engine torque, current gear, etc.) must be somehow interpolated to reach a valid state in the simulation model.

Additionally, simulating using the nanoscopic traffic simulator naturally requires a high-resolution representation of the simulated area, especially the road network, but furthermore a 3D model of the environment if this is necessary for the employed sensor models. Microscopic simulation usually works with rather coarse road networks, which are available publicly from sources like OpenStreetMap. Nanoscopic simulation, however, has additional requirements, e.g., the continuity between road segments due to the vehicle dynamic simulation, which need to be satisfied. Obtaining the data basis in the necessary detail is not always directly possible and can cost an additional amount of time and money.

7 Conclusion

In this paper, we proposed a concept for coupling traffic simulators of different simulation resolutions to achieve a multi-resolution traffic simulation which focuses on a dynamically determined area of interest. The presented methodology partitions the simulation area into a variable, highly detailed region of interest represented by a nanoscopic model and the surrounding area simulated at low resolution by a microscopic model. The evaluation shows a dramatic reduction of computation time in comparison with a pure nanoscopic simulation of the same simulation dimensions, which even makes real-time simulation possible. This divide-and-conquer strategy enables accurate, realistic, and large-scale testing and validation of real implementations of driver assistance systems based on vehicular networks in a virtual environment. As the next steps, we are investigating the application of the multi-resolution simulation methodology for the other domains relevant for the simulation of vehicular networks, namely network simulation and application emulation, to model the whole system across all domains efficiently at high fidelity.

References

1. Gietelink O, Ploeg J, De Schutter B, Verhaegen M (2006) Development of advanced driver assistance systems with vehicle hardware-in-the-loop simulations. Veh Syst Dyn 44(7):569–590
2. Hou Y, Zhao Y, Hulme KF, Huang S, Yang Y, Sadek AW, Qiao C (2014) An integrated traffic-driving simulation framework: design, implementation, and validation. Transp Res Part C: Emerging Technol 45:138–153. issn 0968-090X
3. Olstam J (2009) Simulation of surrounding vehicles in driving simulators, PhD thesis, Linköpings universitet, Kommunikations- och transportsystem. In: Linköping Studies in Science and Technology. Dissertations, no. 1248
4. Naus G, Vugts R, Ploeg J, van de Molengraft M, Steinbuch M (2010) String-stable cacc design and experimental validation: A frequency-domain approach. IEEE Trans Veh Technol 59(9):4268–4279
5. Hoogendoorn SP, Bovy PH (2001) State-of-the-art of vehicular traffic flow modelling. Proc Inst Mech Eng Part I J Syst Control Eng 215(4):283–303
6. Ni D (2006) A framework for new generation transportation simulation. In: Proceedings of the 38th conference on winter simulation, ser. WSC '06. plus 0.5em minus 0.4em winter simulation conference, pp 1508–1514
7. Sommer C, German R, Dressler F (2011) Bidirectionally coupled network and road traffic simulation for improved IVC analysis. IEEE Trans Mob Comput 10(1):3–15
8. Ros FJ, Martinez JA, Ruiz PM (2014) A survey on modeling and simulation of vehicular networks: communications, mobility, and tools. Comput Commun 43:1–15
9. Booysen MJ, Zeadally S, Van Rooyen G-J (2013) Impact of neighbor awareness at the MAC layer in a vehicular ad-hoc network (vanet). In: 2013 IEEE 5th international symposium on wireless vehicular communications (WiVeC), plus 0.5em minus 0.4em. IEEE, New York, pp 1–5
10. Sommer C, Krul R, German R, Dressler F (2010) Emissions versus travel time: simulative evaluation of the environmental impact of its. In: 71st IEEE vehicular technology conference (VTC2010-Spring). plus 0.5em minus 0.4em. Taipei, Taiwan: IEEE, pp 1–5
11. Davis PK, Hillestad R (1993) Families of models that cross levels of resolution: Issues for design, calibration and management. In: Proceedings of the 25th conference on winter simulation. plus 0.5em minus 0.4em. ACM, pp 1003–1012
12. Flötteröd G, Nagel K (2007) High speed combined micro/macro simulation of traffic flow. In: IEEE intelligent transportation systems conference, ITSC, pp 1114–1119
13. Rieck D, Schünemann B, Radusch I, Meinel C (2010) Efficient traffic simulator coupling in a distributed v2x simulation environment. In: Proceedings of the 3rd international ICST conference on simulation tools and techniques, ser. SIMUTools '10. plus 0.5em minus 0.4em. ICST, Brussels, Belgium, Belgium: ICST (Institute for Computer Sciences, Social-Informatics and Telecommunications Engineering), pp 72:1–72:9
14. Potuzak T (2013) Issues of parallel hybrid nanoscopic/microscopic road traffic simulation. In: EUROCON. IEEE, New York, pp 614–621
15. Navarro L, Flacher F, Corruble V (2011) Dynamic level of detail for large scale agent-based urban simulations. In: The 10th international conference on autonomous agents and multiagent systems - Volume 2, ser. AAMAS '11. plus 0.5em minus 0.4em. Richland, SC: International Foundation for Autonomous Agents and Multiagent Systems, pp 701–708
16. Olstam J, Elyasi-Pour R (2013) Combining traffic and vehicle simulation for enhanced evaluations of powertrain related adas for trucks,. In: 2013 16th international IEEE conference on intelligent transportation systems - (ITSC), pp 851–856
17. Pereira JLF, Rossetti RJF (2012) An integrated architecture for autonomous vehicles simulation. In: SAC, Ossowski S, Lecca P (eds) plus 0.5em minus 0.4em. ACM, pp 286–292
18. Krajzewicz D, Erdmann J, Behrisch M, Bieker L (2012) Recent development and applications of SUMO - Simulation of Urban MObility. Int J Adv Syst Meas 5(3&4):128–138

19. Krajzewicz D (2013) Summary on publications citing SUMO, 2002–2012. In: 1st SUMO user conference - SUMO 2013. DLR. SUMO2013 - 1st SUMO user conference, 2013, Berlin, Germany. ISSN 1866-721X
20. Krauss S, Wagner P, Gawron C (1997) Metastable states in a microscopic model of traffic flow. Phys Rev E 55:5597–5602
21. von Neumann-Cosel K, Dupuis M, Weiss C (2009) Virtual test drive - provision of a consistent tool-set for [d,h,s,v]-in-the-loop. In: Proceedings of the driving simulation conference Monaco
22. Treiber M, Hennecke A, Helbing D (2000) Congested traffic states in empirical observations and microscopic simulations. Phys Rev E 62(2):1805
23. Wegener A, Piórkowski M, Raya M, Hellbrück H, Fischer S, Hubaux J-P (2008) Traci: an interface for coupling road traffic and network simulators. In: Proceedings of the 11th communications and networking simulation symposium, ser. CNS '08. plus 0.5em minus 0.4em. ACM, New York, pp 155–163
24. Reynolds PF Jr, Natrajan A, Srinivasan S (1997) Consistency maintenance in multiresolution simulation. ACM Trans. Model. Comput. Simul. 7(3):368–392
25. Krauß S (1998) Microscopic modeling of traffic flow: investigation of collision free vehicle dynamics, PhD thesis

How Does the Traffic Behavior Change by Using In-Vehicle Signage for Speed Limits in Urban Areas?

Laura Bieker

Abstract Speeding is a major risk factor for accidents. Reminding the driver of the current speed limit might improve traffic safety. The impact of In-Vehicle Signage (IVS) for speed limits via Car2Infrastructure communication was investigated in a test field in Tampere (Finland). The test field results show that IVS has a positive influence on the speed of the drivers of the equipped vehicle. Different scenarios for the estimated penetration rates of equipped vehicles were set up to see the overall effects on the traffic efficiency by IVS. The effects on the traffic efficiency were simulated in SUMO. The simulation results show that no benefit in traffic efficiency for all traffic participants could be reached by IVS.

Keywords C2X · In-vehicle signage · Traffic efficiency

1 Introduction

Speeding of vehicle drivers is a common problem in traffic safety. Studies indicated that speeding is one of the most contributing factors to car accidents. Speeding was the main reason for an accident in 30% of all accidents in the USA in 2012 [1]. The percentage of speeding-related incidents is rising. There are different reasons why drivers are violating speed limits. Many of speeding accidents have recorded the misuse of alcohol (according to [1] 42%). But speed limits can also be easily overlooked especially from younger drivers who have not much driving experience. Therefore, an application called In-Vehicle Signage was developed to remind the driver of the current speed limit. The main idea was that many drivers would reduce their speeds if they would be more aware of the current speed limit. An implication of the reduced speed could be an effect on traffic safety and/or traffic efficiency.

Other studies indicate that speed control algorithm can also have an influence in traffic efficiency. In [2], it is described how jam waves can be reduced by a cooperative system for variable speed limits. The described algorithm uses floating car data

L. Bieker (✉)
German Aerospace Center, Rutherfordstraße 2, 12489 Berlin, Germany
e-mail: Laura.Bieker@DLR.de

© Springer International Publishing AG, part of Springer Nature 2019   37
M. Behrisch and M. Weber (eds.), *Simulating Urban Traffic Scenarios*,
Lecture Notes in Mobility, https://doi.org/10.1007/978-3-319-33616-9_3

and video-based monitoring to detect jams and resolve the jams via variable speed signs. The study in [3] presents an approach to optimize the vehicular traffic flow on highways. The traffic flow could be improved in this approach by presenting an individual speed limit for each traffic participant. The studies show that driver depended speed recommendation can have an influence on traffic flow and jams on highways. Furthermore, this paper investigates whether speed recommendation can also have an influence on traffic efficiency in urban areas.

First, this paper gives an overview of the evaluated application called In-Vehicle Signage for speed limits. Next, the test field and the measured data are described. Afterward, the used simulation scenario and results are presented. Finally, concluding remarks and an outlook is given.

2   In-Vehicle Signage for Speed Limits

The purpose of the In-Vehicle Signage (IVS) function is to display the traffic signs on an onboard unit inside the vehicle to improve the driver's perception of the sign.

The IVS function is implemented via Car2Infrastructure communication (C2I). This means that the car is sending and receiving messages from/to a roadside unit (RSU) along the road. In this study, the IVS for speed limits in urban areas was investigated. The onboard unit reminds the driver of speed limits and warns if he/she is violating it. The visual warning is given at the location of the traffic sign and was visible for 100–200 m (depending on the relevance and physical environment).

IVS informs the driver under the following circumstances: when driving just under (90–100%), just above (100–110%) or significantly above (over 110%) the speed limit. Here, the purpose of this system is to make the driver aware of how he/she is driving with respect to the speed limit. The purpose of the IVS is to encourage the driver to slow down when exceeding the speed limit. The IVS application was tested in a real-world test field within the DRIVE C2X project.

3   Test Field in Tampere (Finland)

For this study, the test field data from the DRIVE C2X project was used. In the EU project DRIVE C2X, seven test sides have been built up [4]. The aim of the project was to build a foundation for cooperative vehicle system in Europe. A basic set for C2X services has been tested at these locations; see Fig. 1. The set of applications includes:

- Traffic Jam Ahead Warning
- Green Light Optimal Speed Advisory
- In-Vehicle Signage
- Weather Warning

Fig. 1 Test sides of DRIVE C2X [4]

- Road Works' Warning

For this study, the data from the test field in Tampere (Finland) was used; see Fig. 2. The test field has roads with different speed limits. VTT was operating the test field and was supported by the city of Tampere. The test field includes 8 km of urban roads and with a traffic demand of normally 1000–2500 vehicles per hour.

4 Results of the Test Field

The test drives in Tampere were performed from April 24 to May 08, 2013. The cars of the test drivers were equipped with Car2Infrastracture communication (C2I). Along

Fig. 2 Finnish test field

the streets, four RSUs were installed to collect the data of the equipped vehicles and send information about the traffic signs. Two test drive scenarios exist:

- Baseline scenario: The vehicle is driving like a normal vehicle without IVS. Only the vehicle data is collected via C2I.
- Treatment scenario: The vehicle is driving with IVS.

The dataset includes 544 events, with equally 277 for baseline and treatment events. The speed data of the test field was analyzed for the different speed segments within the test field. Figure 3 shows the average speeds for the whole test drive, the roads with a speed limit of 30 km/h and of 40 km/h. In addition, the speed data was used to analyze whether the test field drivers regard the speed limits. The result was that 26.8% of the drivers drive faster than the speed limits if they drive without the IVS application but only 17.8% of the driver with the IVS application.

The analyzed data shows that the speed limit warnings have a small positive effect on the driving behavior in terms of reduction of speed (of 4% at 30 km/h

Fig. 3 Average speed of the different speed segments within the test field

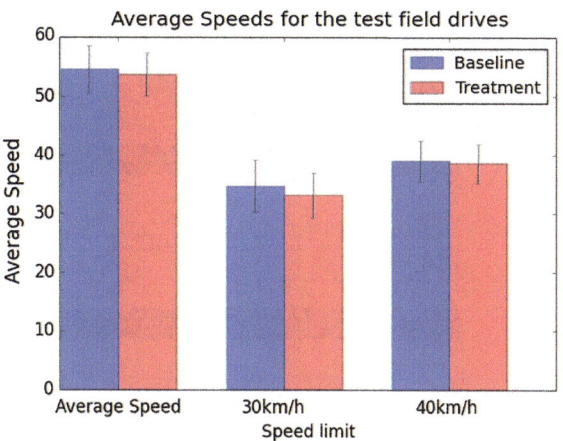

speed limit and of 1% at 40 km/h speed limit). A reduced speed also implies often reduced fuel consumption and produced emissions. These positive results are only for the equipped vehicles, but it would be interesting to see how these results could influence the traffic behavior overall.

5 Simulation

The test field data indicates the impact of the IVS application for a single vehicle. To scale up the results of the equipped vehicles on a level of a whole vehicle population (including equipped and non-equipped vehicles), a simulation in SUMO was performed [5].

5.1 Simulation Scenario

An urban road of 1 km length was used as simulation scenario; see Fig. 4. The same scenario is simulated with a speed limit of 30 km/h and 40 km/h to analyze the difference on traffic efficiency in both cases. A road with two lanes was chosen so that all vehicles are driving in the same direction. The opposing traffic is assumed to have no effects on the driving behavior. Simulation runs with only one lane showed that the vehicles have to adapt their speed to the leading vehicle with the lowest speed because no overtaking is possible.

IVS is not expected to have an effect on route choice. Hence, it is sufficient to analyze corridor networks without route alternatives. In SUMO, every driver has an own speed which is randomly computed for every street and the corresponding speed limit. For every vehicle type, a speed factor and a standard deviation can be given. The

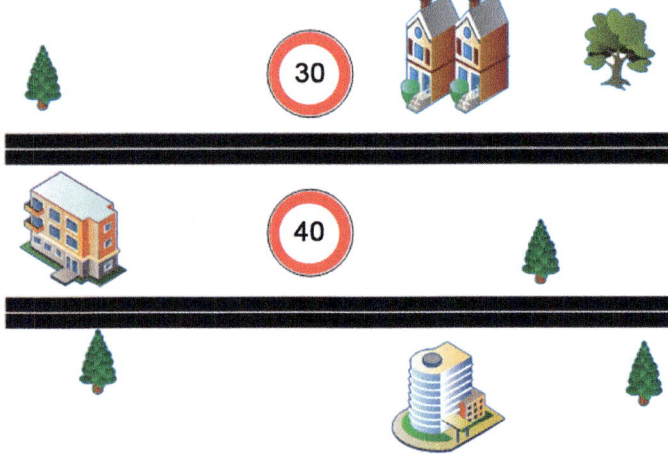

Fig. 4 Urban scenario with speed limits 30 and 40 km/h, two lanes in the same direction

desired speed of each vehicle is randomly given according to a lognormal distribution with the average speed and the standard deviation from the field test. Additionally, the standard deviation is also given for the field test data. Given these figures, a lognormal distribution can be calculated and the desired speed of the vehicles is set according to the speed distribution 100 m in front of the speed limit, so the vehicle has time to adapt its speed.

5.2 Penetration Rate

For the simulation, different scenarios of penetration rates of the vehicle within the simulation were investigated within the DRIVE C2X project [6]:

- a main estimate,
- a pessimistic estimate, and
- an optimistic estimate.

Figure 5 shows graphs of these estimates over time. The estimation of penetration rate used different factors for the automobile market including, e.g., number of new cars every year and percentage of vehicles which will probably have implement new technologies and applications (Fig. 5).

It was not feasible to use all penetration rates for the simulation. Hence, four cases were chosen to be used for the DRIVE C2X project. The four cases can be seen in Table 1. It was needed to have one "baseline" scenario to compare the other simulation results with, so the year 2010 without equipped vehicles was chosen. Additionally, it should be investigated whether low penetration rates have a significant impact on the traffic; therefore, a "low" scenario was chosen.

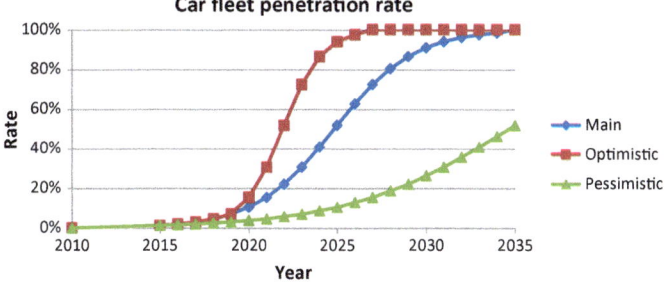

Fig. 5 Fleet penetration rate estimates for all DRIVE C2X systems [6]

Table 1 Fleet penetration rate

Scenario	Car penetration rate (%)
Baseline (2010)	0.00
Low (2020, Main)	10.39
Medium (2030, Pessimistic)	26.29
High (2030, Main)	90.84%

Every simulation scenario was run 40 times with a high traffic demand (peak) and low traffic demand (off-peak). The results of the simulation runs can be found in the next section.

6 Results

The travel time and the delay time were calculated for the whole trip of each vehicle within the simulation. The results for the average travel time of the vehicles can be seen in Fig. 6. For each scenario, the average travel time is around 94 s, and there are no significant differences. Likewise, the average delay is around 17 s and is only slightly changing for different equipment rates (Fig. 7).

For analyzing the results of the simulation, three induction loops were included within the scenario to measure the local mean speed. The first speed detector was 200 m before the speed limit sign, one was directly at the location of the speed limit sign and one was 200 m after the speed limit sign. The result can be seen in Fig. 8. At the first induction loop, the average speed is around 50 km/h and vehicles are all adapting their speed to the current speed limit at the second and third induction loop.

Fig. 6 Average travel time
of the vehicles within the
different simulation
scenarios

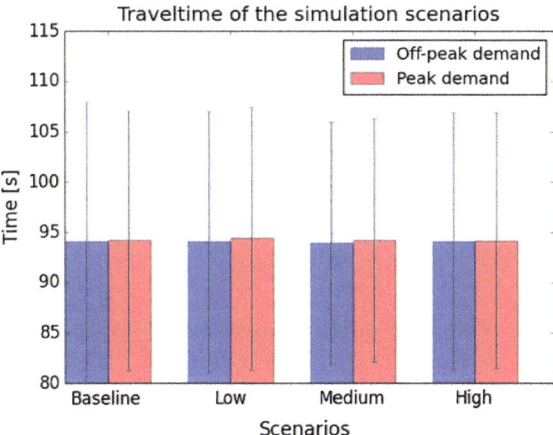

Fig. 7 Average delay of the
vehicles in the simulation
scenarios

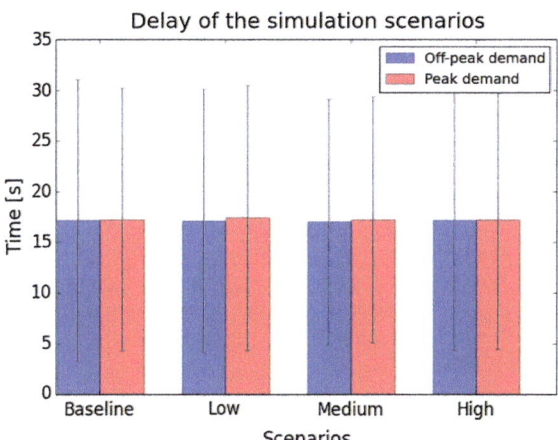

7   Discussion

It was expected that the travel time and the delay would increase with higher equipment rates and the average speed at the induction loops will decrease. But the differences in travel time, delay, and speed are in almost all scenarios very small and
not significant compared to the baseline scenario. An explanation could be that the
measured speed differences in the test fields are too small to see the difference in a
stochastic simulation. Namely, the changes are 1.507 km/h for speed limit 30 km/h
and 0.449 km/h for speed limit 40 km/h. These small changes seem to be neglected
in the simulation with many traffic participants.

One way would be to check whether significant results could be produced if the
speed difference for equipped and non-equipped vehicles would be larger. Therefore,
the same simulation scenario was set up with artificial values for the speed adaptation.

Fig. 8 Average speeds with a low traffic demand at the location of the induction loops (200 m before the speed limit, at the location of the speed limit and 200 m after the speed limit)

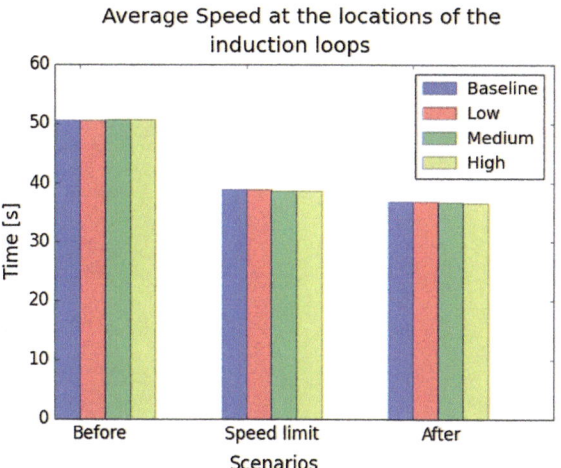

Fig. 9 Travel time result for an artificial example

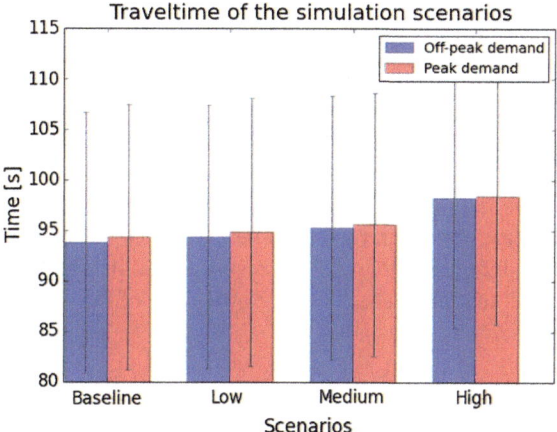

The same values for the average speed of the non-equipped vehicles (34 km/h at a 30 km/h speed limit and 39 km/h for the 40 km/h speed limit) were used while for the equipped vehicles the average speed adopted to 30 km/h at a 30 km/h speed limit and 35 km/h at a 40 km/h speed limit. These values are not realistic and should not be used in other studies; they are only used to validate the algorithm.

The simulation for this artificial simulation scenario can be seen in Figs. 9 and 10. The average travel time is increasing with higher penetration rates of equipped vehicles as it was expected. Also, the average speeds are decreasing for simulation scenarios with higher equipment rates of the IVS application, like it was expected. The simulation results show that the algorithm is working like it was expected but the measured speed changes in the test field are too small to have an impact for the whole traffic.

Fig. 10 Average speeds for
an artificial simulation

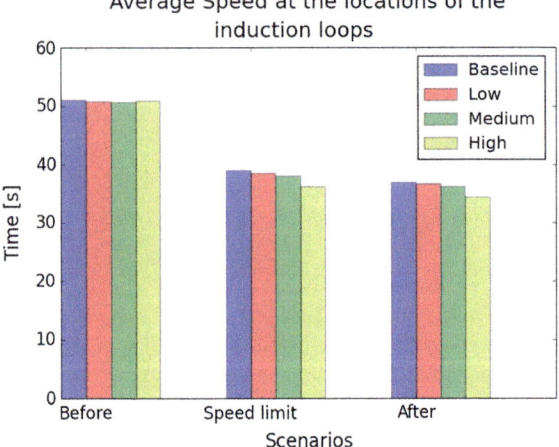

8 Conclusion

The results of the test field indicate that IVS has positive effects on the speed of
equipped vehicles. IVS can influence traffic efficiency by displaying speed sign
information in car and thereby adapting the drivers' desired speed. The effects found
in the field trial were used to model the change in desired speed in the traffic simulations: For urban roads, the desired speeds were lowered by 1–4%. The results are
not significant. It could be presumed that the influence of IVS in urban areas is too
small to see benefits in traffic efficiency of the whole traffic.

References

1. National Highway Traffic Safety Administration: Traffic Safety Facts: 2012 Data. http://www-
   nrd.nhtsa.dot.gov/Pubs/812021.pdf. Accessed 25 June 2015
2. Hegyi A, Netten BD, Wang M, Schakel WJ, Schreiter T, Yuan Y, Arem B, van Alkim, T (2013)
   A cooperative systems based variable speed limit control algorithm against jam waves - an
   extension of the SPECIALIST algorithm. Proceedings of the 16th international IEEE annual
   conference on intelligent transportation systems (ITSC 2013),The Hague, pp 973–978, Oct 6–9,
   2013
3. Forster M, Frank R, Engel T (2013) A study on highway traffic flow optimization using partial
   velocity synchronization. In: Proceedings of the 1st GI/ITG KuVS fachgespräch inter-vehicle
   communication (FG-IVC 2013) technical report CCS-2013-01, 19–22
4. DRIVE C2X (2015) http://www.drive-c2x.eu/test-sites. Accessed 23. Jan 2015
5. Krajzewicz D, Erdmann J, Behrisch M, Bieker L (2012) Recent development and applications
   of SUMO - simulation of urban mobility. Int J Adv Syst Meas 5(3, 4):128–138
6. DRIVE C2X Deliverable D 11.4 Impacts of cooperative systems and user perception. http:/
   /www.drive-c2x.eu/tl_files/publications/Deliverables%20and%20abstracts/DRIVE%20C2X_
   D11.4_Impact_Assessment_v1.0_full%20version.pdf. Accessed 23 Jan 2015

Connecting Macroscopic and Microscopic Traffic Assignment

Michael Behrisch

Abstract There are several methods of traffic assignment related to static assignment of link costs and a dynamic assignment based on microscopic traffic simulation which are implemented in SUMO. This paper compares the different approaches and implementations and shows how to combine them to derive a good starting solution for the more precise microscopic approach from the coarse macroscopic solution. In addition, a new tool from the SUMO suite and some first results of applying the schema to the city of Berlin are presented.

Keywords SUMO · Macroscopic assignment · Dynamic user assignment Microscopic traffic simulation

1 Introduction

In the context of the VEU project, a coupling between an agent-based traffic demand model (TAPAS [1]) and a microscopic traffic simulation (SUMO [2]) is being developed. The objective of this connection is to give realistic feedback about the expected travel times to the agents of demand model. This allows for an iterative refinement of the planned activities as well as the expected mobility costs in terms of time but also emission. Since both models are based on the simulation of individual agents, which is a potentially time-consuming task, and they should be applied to scenarios of larger cities like Berlin (3.5 million inhabitants), improving run-time performance is almost as important an issue as the quality of the solution. This paper will focus on improvements over the classical DUA approach to traffic assignment which were implemented in SUMO recently. This covers macroscopic assignment tools as well as combined approaches which should reduce run-time while retaining or even improving the quality of the user assignment.

The next section describes the current state of dynamic user assignment in SUMO together with some drawbacks and limitations, especially concerning large scenarios.

M. Behrisch (✉)
German Aerospace Center, Rutherfordstraße 2, 12489 Berlin, Germany
e-mail: Michael.Behrisch@DLR.de

Section 3 gives some details on fast alternative microscopic route finding approaches using only a single simulation run. Section 4 describes the new MARouter tool recently introduced into the SUMO suite and how it can be applied to solve some of the limitations of the pure microscopic approach. Sections 5 and 6 describe the test scenarios and give some first results of the approach.

2 Microscopic Traffic Assignment with SUMO

The current microsimulation approach in SUMO involves the repeated execution of a routing step and a detailed microsimulation on the resulting routes leading to a new traffic situation. This traffic situation is measured edgeways mostly based on the single measurement of travel time but sometimes also on other measurements such as fuel consumption [3]. The resulting edge costs are fed back into a router which calculates new route costs and probabilities (based on a model by Gawron [4]) trying to achieve a user equilibrium state, where no user can reduce its individual route cost by changing to a different route. As the description already indicates this process involves a large number of microscopic traffic simulations which in itself is already a time-consuming task (simulating a city like Berlin for a whole day can easily take up to several hours). Usually, there are constraints applied to limit the number of iterations either to a fixed amount or to some limit in the variation of the travel times among the available routes for each car.

Still, there are major complications in using this approach because it tends to give unstable results for large scenarios where after a relative period of stability later iterations show very different travel times (deviating in both directions). Furthermore, it tends to recover only very slowly from iterations in which the network was completely jammed. For our scenario, we took the most basic approach, evaluating an assignment as being successful, when the number of cars being locked (a car is considered locked when it does not move for more than five minutes) is below 5% of the total number of cars. In the conventional microscopic approach, we failed to fulfil this criterion using the Berlin scenario described below, so we needed to plug in another method leading to two different strategies, one being the MARouter approach described in Sect. 4. The other strategy is based on direct routing in the simulation and will be described in the next section.

The traffic assignment process itself is still subject to a number of improvements since it can, for instance, easily get confused if there are a lot of routes between an origin and a destination travelling the same jammed edge, because it stores by default only travel times for (a limited number of) whole routes and not for single edges. It is also expected that a macroscopic preprocessing can help in avoiding an early breakdown of the iterative assignment process due to such events.

3   One-Shot Routing

SUMO allows for different ways to run a simulation without complete a priori route information. The easiest way is to give the inputs which can be given to the DUARouter tool directly to the simulation in the form of trips or flows which only define a source and a destination edge. The simulation will then find a fastest route based on the current state of the traffic network; that is, it will determine the current travel times on all edges and find a minimum cost path using the same algorithm (usually Dijkstra or A* which can be selected by the user) as the DUARouter. It can also be configured whether routing should happen only on the departure of the vehicle or on a fixed interval during the run-time of the vehicle as well.

The whole process is not only possible for vehicles which have a defined start and end edge but also for trips defined between districts and traffic assignment zones. Since here the departure edge may change, it is possible to define another interval for periodic rechecking whether a new departure (and possibly arrival) edge should be chosen

There are several possibilities to tweak this process in terms of quality, sensitivity and run-time performance. The most obvious parameter is the frequency of the rerouting which can be increased to have less routing overhead but also less sensitivity to upcoming traffic jams. On the other hand, choosing small values can lead to oscillation, especially when the scenario contains traffic lights and the vehicles try to avoid every red traffic light. A good intermediate value we experienced is 300 s.

The second set of parameters affects the determination of the edge weights. It is possible to define the interval with which they are updated (default one second, we used 10) as well as smoothing method and parameter. Per default, the simulation uses an exponential smoothing method where the new weight for an edge is calculated by multiplying the old value with a parameter b and the new value with (1-b) and adding the result. In our runs, we used the default value of 0.5. SUMO can also use a moving window averaging here. Last but not least, the values for the edges can be loaded from a pre-calculated table with edge weights. This speeds up the initialization phase where the net slowly fills up and not all edges already have realistic travel times. This method can be used to couple the MARouter results described below with the one-shot routing.

If the simulating computer can make use of several processor cores, it is possible to speed up the routing process of large scenarios by using multiple threads (we used 16 in our case, leading to a fourfold speedup of the routing). It has to be noted, however, that the speedup is only visible in scenarios where enough routing queries happen at the same time so either many vehicles departure are in the network or rerouting happens very often. Without parallelization, the routing can easily consume more than 90% of the simulation time and it is therefore always advisable to speed it up as much as possible.

4 The Macroscopic Assignment Tool (MARouter)

The MARouter (short for macroscopic assignment router) is a tool which was introduced into the SUMO suite starting with the release of version 0.22.0. It implements standard algorithms for macroscopic traffic assignment such as the iterative assignment algorithm and the stochastic user assignment based on Lohse.

The input for a macroscopic assignment run is either an origin–destination matrix (OD matrix) containing aggregated information about trip counts between traffic zones or a list of trips which is then aggregated internally but disaggregated before output. The routing is currently available for individual cars only so there is no public transport involved.

The MARouter is (for large scenarios) much faster than the iterative microscopic assignment since it replaces the microscopic simulation step by a simple calculation of expected travel times based on the measured demand. This allows for a fast iteration but loses all the benefits of microscopic simulation such as detailed junction modelling and interaction of traffic flows. In order to keep the microscopic advantages in the result, we coupled the approaches to start with a macroscopic assignment and then feed the results into the microscopic loop mentioned above.

4.1 Macroscopic Traffic Assignment

The MARouter implements a traditional static approach to traffic assignment. The input consists of an input area (a SUMO network) subdivided into so-called traffic analysis zones (TAZs) and an origin–destination matrix (OD matrix) which is given for a fixed time interval (usually a day) and a mode of transport (e.g. passenger cars). This matrix has one entry for each pair of origin A and destination B defining the amount of vehicles which move (or rather start moving) from A to B. A second input into the model is a set of capacity restraint (CR) functions which model how the travel time on an edge increases with the volume of traffic on that edge. The MARouter currently uses functions of the following form:

$$t(x) = t_0 * \left(1 + c * \frac{x}{l}\right)$$

where t_0 is the travel time in the empty network, l is the number of lanes, and c is a constant depending on features like the road class (highway or urban road) and the maximum allowed speed. At the moment, those functions are hard coded into the SUMO codebase and cannot be altered by the user (except by modifying the source code). They can be found in the file src/marouter/ROMAAssignments.cpp. The MARouter calculates from those OD matrices and functions an assignment which is a set of routes for each OD pair together with a number of vehicles driving this route. In order to do so, several basic algorithms can be employed which calculate those mappings iteratively. The most basic approach called iterative assignment assigns a fixed percentage of the demand to the net, recalculates the expected travel times given the CR functions above and assigns the next part.

The second implemented approach is the stochastic user equilibrium (SUE). This is much more similar to the microscopic approach in that it calculates multiple routes for every driver where it can choose from.

For further algorithms, we refer to the excellent descriptions in [5].

4.2 Input Formats

SUMO accepts OD matrices in different formats which originate either from VISUM or from the AMITRAN project [6]. An OD matrix in the standard VISUM format looks like the following:

```
$V
* From-Time  To-Time
0.00 24.00
* Factor
1.00
*
* some
* additional
* comments
* District number
2
* names:
         1              2
*
* District 1 Sum = 100
         0            100
* District 2 Sum = 0
         0              0
```

All lines starting with asterisks are comments so the important information is the time span in the beginning, the number and names of the TAZ (districts) and then the real demand given matrix line by line. In the given example, 100 vehicles drive from "1" to "2" during a whole day. In the AMITRAN format, the same input looks like this:

```
<?xml version="1.0" encoding="UTF-8"?>
<demand xsi:noNamespaceSchemaLocation="http://sumo.dlr.de/xsd/amitran/od.xsd">
  <actorConfig id="0">
    <timeSlice duration="86400000" startTime="0">
      <odPair amount="100" destination="2" origin="1"/>
    </timeSlice>
  </actorConfig>
</demand>
```

In addition to the VISUM format, the AMITRAN format contains information about the mode (or rather the vehicle class) it describes as well. This is encoded in the actorConfig which refers to an existing vehicle type for SUMO. Since the format only allows for numeric ids, the mapping from the actorConfig to the vehicle type is done in a separate input file.

4.3   Output Format

The MARouter creates SUMO route files in the standard SUMO format describing the routes and the percentages of usage between the different OD relations. For the example given above, the output would look like this:

```
<?xml version="1.0" encoding="UTF-8"?>
<routes xsi:noNamespaceSchemaLocation="http://sumo.dlr.de/xsd/routes_file.xsd">
    <routeDistribution id="0">
        <route cost="90.24" probability="100.00000000" edges="middle end"/>
    </routeDistribution>
</routes>
```

MARouter can also directly output flows which can be fed into the simulation or the DUARouter/dualterate.py procedure:

```
<?xml version="1.0" encoding="UTF-8"?>
<routes xsi:noNamespaceSchemaLocation="http://sumo.dlr.de/xsd/routes_file.xsd">
    <flow id="0" begin="0.00" end="86400.00" number="100" fromTaz="1" toTaz="2">
        <routeDistribution>
            <route cost="90.24" probability="100.00000000" edges="middle end"/>
        </routeDistribution>
    </flow>
</routes>
```

In this example, MARouter determined only a single route between the source and the destination. The probability in the described output is not normed to 1 in order to allow for easy recalculation of the number of vehicles travelling each of the routes.

5   The Berlin Scenario

The test scenario is the individual traffic demand (only passenger cars, no public transport and no delivery/goods traffic) for the city of Berlin consisting of approximately three million trips a day between 1100 traffic zones. The TAPAS model sets up a virtual population for the city of Berlin and determines its transportation demands for an average weekday. Each person has an assigned activity plan representing destinations like work and school places and the time they spent there. The input data is based on various sources mainly statistical data from the Berlin Microcensus and the "Mobilität in Deutschland" study. For a detailed description, see [1]. The model generates trips for all available modes of transportation, but in the current implementation only the individual traffic is used for the later microsimulation. The input from the TAPAS simulation consisting of individual trips was aggregated on the traffic zone level for running the MARouter. The resulting edge loads are fed into the microscopic dynamic user assignment procedure as described in Sect. 2 (Fig. 1).

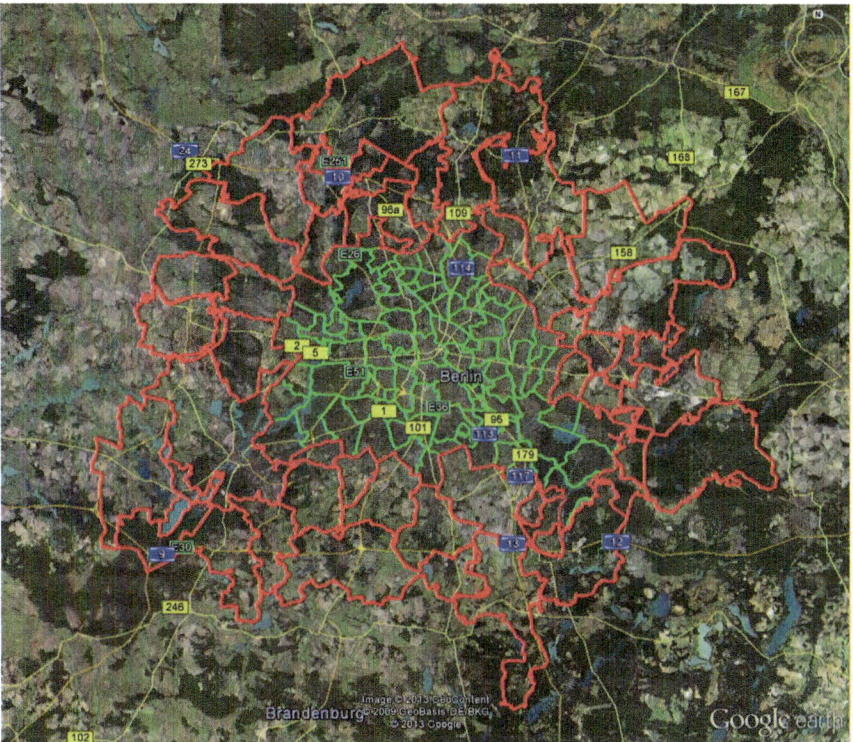

Fig. 1 Districts of the city of Berlin together with the surroundings (courtesy Google Earth). Each of the depicted districts corresponds to roughly 20 traffic analysis zones

Afterwards, a comparison between the raw microscopic process and the process with macroscopic preprocessing could be performed. Unfortunately, the whole process still takes in the order of a day to complete so that there are only preliminary results available yet.

6 Conclusion and Discussion

The first results show a speedup of the user assignment runs. To reach a comparable result, the run-time decreases in a range of 15–20%. From the small sample set, it is hard, however, to judge the final results, especially concerning the quality of the solution. More detailed results are expected in further research and will be presented at the conference.

While the speedups are currently not as high as expected, the MARouter tool itself seems to be of high value because now for the first time we can directly compare macroscopic and microscopic assignment strategies on the same data set. Further results also on abstract networks will be presented in an extended version of this paper.

References

1. Justen A, Cyganski R (2008) Decision-making by microscopic demand modeling: a case study. In: Transportation decision making: issues, tools, models and case studies. genesidesign.com. Transportation decision making: issues, tools and case studies, 2008–11-13 - 2008-11-14, Venedig
2. Behrisch M, Bieker L, Erdmann J, Krajzewicz D (2011) SUMO—simulation of urban mobility: an overview. In: SIMUL 2011, the third international conference on advances in system simulation
3. Flötteröd Y-P, Wagner P, Behrisch M, Krajzewicz D (2012) Simulated-based validity analysis of ecological user equilibrium. In: Winter simulation conference archive. 2012 winter simulation conference, 09–12 Dec 2012. Berlin, Deutschland
4. Gawron C (1998) Simulation-based traffic assignment – computing user equilibria in large street networks. PhD Dissertation, University of Köln, Germany
5. Lohse D, Schnabel W (2011) Grundlagen der Straßenverkehrstechnik und der Verkehrsplanung: Band 2 – Verkehrsplanung, Beuth Verlag
6. AMITRAN consortium (2014) AMITRAN web pages, http://www.amitran.eu/. Accessed 10 April 2015

A SUMO Extension for Norm-Based Traffic Control Systems

Jetze Baumfalk, Mehdi Dastani, Barend Poot and Bas Testerink

Abstract Autonomous vehicles will most likely participate in traffic in the near future. The advent of autonomous vehicles allows us to explore innovative ideas for traffic control such as norm-based traffic control. A norm is a violable rule that describes correct behavior. Norm-based traffic controllers monitor traffic and effectuate sanctions in case vehicles violate norms. In this paper, we present an extension of SUMO that enables the user to apply norm-based traffic controllers to traffic simulations. In our extension, named TrafficMAS, vehicles are capable of making an autonomous decision on whether to comply with norms. We provide a description of the extension, a summary on its implementation and demonstrative experiments.

1 Introduction

Recent developments in the automotive industry steer toward a future where autonomous vehicles are part of everyday traffic (cf. [1, 2, 15]). Vehicles will no longer have a human driver, but will drive themselves and communicate with smart road infrastructures and other vehicles. Clearly, human drivers are different from software programs that operate vehicles. For instance, the response of human drivers to receiving events is considerably slower and less accurate in comparison with a computer program. These differences pose new challenges and opportunities [8]. For example, delegating cruise control to the vehicles' board computers allows vehicles to coordinate and form platoons, which improves the traffic flow [12].

J. Baumfalk (✉) · M. Dastani · B. Poot · B. Testerink
Utrecht University, Utrecht, Netherlands
e-mail: J.T.Baumfalk@uu.nl

M. Dastani
e-mail: M.M.Dastani@uu.nl

B. Poot
e-mail: B.W.A.Poot@uu.nl

B. Testerink
e-mail: B.J.G.Testerink@uu.nl

© Springer International Publishing AG, part of Springer Nature 2019
M. Behrisch and M. Weber (eds.), *Simulating Urban Traffic Scenarios*,
Lecture Notes in Mobility, https://doi.org/10.1007/978-3-319-33616-9_5

In this paper, we are particularly interested in future challenges and opportunities for traffic control. Traffic controllers can exert some level of influence on vehicles in order to improve traffic flow and safety. Currently, traffic is controlled by means of traffic laws and signs which require the education of general traffic regulations and the interpretation of signs. The government as a regulator creates incentive to follow the regulations by imposing sanctions on anybody who is caught violating them. This control mechanism is tailored for humans, as they are currently the road's only occupants. For example, speed limits are given in easy round numbers, as we expect humans to only approximate their limits within an error margin. An autonomously controlled vehicle has a more precise control over its velocity and hence its error margins are different. This allows us to give an autonomous vehicle more precise directives. We shall address the question of how we can design, analyze, and test new traffic controllers where (a portion of) the vehicle is driven by autonomous software systems.

We envision a future where the traffic infrastructure consists of smart roads, enriched with software systems that control traffic. We will call such a software system a traffic controller. The function of a traffic controller is to observe and evaluate traffic, and communicate personalized directive to vehicles. Such a traffic controller is required to respect the autonomy of the vehicles as autonomous vehicles are assumed to be self-interested with possibly personal incentives to violate traffic regulations [5]. Traffic controllers will be allowed to impose sanctions on autonomous vehicles as a way to promote desired behavior. Thus, in our vision autonomous vehicles, which are aware of traffic regulations and their corresponding sanctions, may still violate traffic regulations and accept the imposed sanctions whenever their personal objectives are worth the incurred sanctions. Traffic controllers should also be easily maintainable in terms of the traffic regulations. In our approach, we focus on drivers, the state of traffic, the regulations, and the traffic controllers. We shall argue that a suitable paradigm for these kinds of control systems is that of norm-based control for multi-agent systems. Norms are violable regulations, whose violations result in the imposition of sanctions, much like current traffic regulations.

Aside from design, we also want to see what the effect is of a new traffic control system. As real-world experiments are an expensive affair, a traffic control research project starts often with traffic simulations. We choose SUMO [13] as a simulation platform because it is open-source, has a track record of research behind it, and performs well. We did observe, however, that SUMO does not provide a straightforward platform to implement concepts from the paradigm of norm-based control for multi-agent systems. In this paper, we present an extension to SUMO, named TrafficMAS, that allows the user to specify and execute norm-based traffic controllers for traffic. This software is open-source and can be found at https://github.com/baumfalk/TrafficMAS. A small user guide on the extension can be found in Appendix 2, and more information is available on the Github page.

The running example in this paper is a ramp-merging scenario, where two traffic streams have to merge together. For a schematic overview, see Fig. 1. There is one main input stream and one ramp input stream, resulting into a single output stream. The goal is to make optimal use of the output capacity of the network while not

Fig. 1 Example scenario
where two traffic streams
must merge

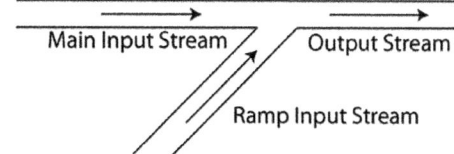

causing unnecessary traffic jams for the ramp input stream or compromising safety. A solid analysis of this scenario can be found in [8]. Our approach is to use a traffic controller that assigns individual directives to vehicles. In particular, vehicles are obligated to move or stay on a lane and/or adopt a certain target speed until they are released of this directive.

Our paper is structured as follows. We first give a brief introduction in the field of norm-based control systems. We will then show how the norm-based control system for SUMO is designed and discuss its application in our example scenario. Following that, we describe how vehicles can reason about the norms that are directed to them by the traffic controller. We then explain our implementation approach. In the next section, we evaluate our extension through a series of experiments that highlight different aspects of our contribution. Finally, we look at related work and compare them with our approach.

2 Norm-Based Control Systems

Norm-based control systems are a popular technology for coordination in the multi-agent systems [10] community. A multi-agent system consists of a set of agents that interact within a shared environment. The agents may communicate with each other or perform actions in their shared environment. In addition to reactively responding to environmental changes, autonomous agents are assumed to have their own objectives (goals) for which they proactively initiate actions in order to achieve them. Aside from its objectives, the knowledge/belief of an agent may determine the actions it decides to perform. A simple model of an agent's internal process is the sense–reason–act cycle. In this cycle, the agent first senses what the state of the environment is; then reasons about its goals, preferences, etc., to determine what action it wants to perform; and then executes the action. Although the agents' behavior is not always predictable or controllable, multi-agent systems are often required to satisfy some global properties. For our purposes, we consider smart roads as a multi-agent system where the state of traffic as well as the infrastructure is seen as the environment, and the drivers of vehicles, whether human or not, are seen as the agents in a traffic MAS. The throughput and safety of smart roads are considered as the global properties that are required to be satisfied.

2.1   Background on Norm-Based Control Systems

The behavior of individual agents needs to be controlled and coordinated in order to ensure the desirable properties at the system level [10]. A possible approach would be to design and implement hard constraints at both individual agent and multi-agent system levels. However, this approach may not always be desirable or feasible since limiting the autonomy of individual agents can be costly or even impossible. For instance, in our smart roads scenario it is undesirable, if not impossible, to fully control and determine the behavior of all individual autonomous cars as these vehicles are assumed and designed to be able to make their own decisions.

Norm-based control systems are widely proposed as an effective mechanism to control and coordinate the behavior of individual agents [9]. In norm-based control systems, norms are considered as specifying the standards of behavior that can be used to govern the interaction between autonomous agents (cf. [4, 11, 19]). Across various theories and frameworks, there is no general consensus on what a norm is. In this paper, we use the term norm as a reference to both norm schemes and norm instances. With the concept norm scheme, we refer to the specification of the circumstances after which a specific agent is obliged to achieve a system state, and the sanction that will be imposed should this obligation not be met before a certain deadline (cf. [19]). With the concept norm instance, we refer to a specific directive and sanction that is in effect for a specific agent. In general, norms can take the form of an obligation, prohibition, or permission, but the scope of this paper concerns only obligation norms. The application of norms in multi-agent systems, called norm-based multi-agent systems, requires continuous monitoring of the behavior of individual agents, evaluation of their behavior with respect to the specified norms, and assurance that norm-violating agents are sanctioned. This approach maintains the agents' autonomy and can still promote desirable behavior. We observe that traffic regulations can be formulated following the normative approach. For instance, one can straightforwardly formulate a speed limit measure in terms of a condition (entering the road), obligation (maintaining maximum velocity), deadline (passing a camera), and sanction (a fine).

Some applications are inherently distributed and require distributed control systems. There are various benefits for distributed control such as increased robustness, parallel processing of data, less communication of data, and modular maintenance [18]. We believe this is also strongly the case for future traffic in smart roads where sensors and traffic controllers are geographically distributed, different sections of road networks are governed by different sets of norms and regulations, and the amount of data generated and processed is big. Therefore, we aim at decentralized traffic control consisting of decentralized monitoring of sensor data as well as distributed enforcement of norms. Our extension allows for a straightforward implementation of individual traffic controllers and their communication. Although a decentralized traffic control mechanism is more robust in the sense that it avoids the problem of single point of failure, it also introduces new issues. In decentralized monitoring, one has to make sure that the sensors are cooperating correctly to detect norm violations in a timely fashion [17].

2.2 Norm-Based Traffic Controllers for Traffic

A norm-based traffic controller monitors vehicle behavior and reacts to it. The monitoring functionality of traffic controller serves two purposes. The first purpose is to detect violations of norms, such as speeding, tailgating, or driving on a priority lane without a permit. If such a violation is detected, then a sanction coupled with this violation will be issued toward the violating vehicle. The second purpose of monitoring is to issue new norms. If the traffic controller continues to observe situations where either the throughput or safety of vehicles declines, then a norm might be issued to improve the situation. This norm might be global, or tailored to a specific vehicle. This allows the traffic controller to be very adaptive to traffic dynamics. Furthermore, the severity of a sanction might be increased if the coupled violation either occurs an excessive amount of times or seems to be the cause of problematic situations. In our framework for norm-based traffic controller, both autonomous vehicles and the traffic controller show adaptive behavior toward the ever-changing traffic flow and enable the system to cope with difficult and dynamic traffic situations.

A norm-based traffic controller on smart roads has a collection of sensors with which it can sense the state of the environment, in this case, the state of traffic and the infrastructure. Examples of these sensors are inductive-loop detectors or cameras that can perform image processing aside from speed measurements. Traffic regulations are formulated as norms. A traffic controller has a set of norm schemes that given sensor data can make norm instances. The norm instances are maintained by the traffic controller for as long as they are applicable. Finally, a traffic controller may have subscribers and subscriptions. In a decentralized setting, traffic controllers can subscribe to each other in order to receive sensor data which they cannot obtain locally. Hence, we obtain a form of decentralized norm-based traffic control.

The norm-based traffic controllers in our extension are modeled as a cyclic process that continuously senses its environment (sense phase), evaluates the norms (reason phase), and imposes sanctions when norms are violated (act phase). In the sense phase, traffic controllers have two ways of observing the environment: (i) Obtain data from the sensors that are placed in the environment, and (ii) obtain data from other traffic controllers (by means of communication) to which the norm-based controller is subscribed. In case of the SUMO simulations, this happens every tick. Immediately after the data is received from the sensors, information is communicated between the traffic controllers. In the reason phase, the traffic controllers apply their information to their norm schemes in order to instantiate norm, if possible at all. Finally, the traffic controllers act on these new norm instances by communicating the directives to the vehicles in order for drivers to adapt their behavior. Furthermore, the traffic controllers act by imposing sanctions on the drivers when they fail to comply to existing obligations when a deadline has been reached.

In our smart road application, norm instances are announced by the norm-based traffic controller to autonomous vehicles, just like drivers are assumed to be aware of traffic regulations. Norm awareness allows an autonomous vehicle to reason and decides whether or not to comply with the norms (cf. [3, 16]). Norm awareness is crucial for future traffic on smart roads as autonomous vehicles need to know the consequences of norm violations before deciding whether or not to comply with the norms.

3 Application of Norm-Based Traffic Control in SUMO

In our use-case scenario, we illustrate a fairly simple norm-based traffic controller. The traffic controller first calculates when individual vehicles would arrive on the merge point of the traffic streams. The algorithm used is described in [21] and returns an ordering of vehicles. Next, the target velocity is calculated for each individual vehicle that ensures that it a) crosses the merge point at least two seconds after the vehicle that will cross the point before it, and b) the maximum safe velocity is still maintained. In case the mainstream road has two lanes, a vehicle can be obliged to move to the left lane too. This happens when target velocity of a vehicle on the main road is below a predefined threshold. The sanctions of violating norms are captured by a low or high fine.

In Fig. 2, a schematic representation of the aforementioned ramp-merging scenario is given. Triangles are vehicles that travel in the direction toward they point. White vehicles are the ones that have not yet received their directive from the traffic controller, while the black vehicles have passed a sensor and have thus received a personalized norm instance. The vehicles without a norm instance in Fig. 2 are vehicles A, B, C, and D. Lane sensors (s_1 to s_5) are placed on the roads. These sensors can detect the status of vehicles that are driving over them. For the scenario to work correctly, it is necessary that either the sensors are sufficiently long or the vehicles are sufficiently slow, so that no vehicle can pass the sensors undetected. The sensors should also be placed at a distance far enough from the merge point m such that vehicles have enough time to comply with the directive before the deadline. There are two important points on the road: point m where the two roads merge, and point e where the vehicles exit the scenario. Distances d_A and d_C are agents A and C's distances to m, d_{exit} is the distance from the m to e, and d_{safe} is the distance between vehicles that are deemed safe (i.e., the minimal gap between cars). Ideally, A and C traverse d_A and d_C such that they arrive at m with a distance d_{safe} and can accelerate to their maximum speed within the distance d_{exit}.

Monitoring happens through interpreting the observations of sensors. In the case of SUMO, we use lane detectors that can sense the vehicles that driving along the area they cover. Specifically, each sensor can detect the identity, velocity, and position of each vehicle on the sensor's area. Further parameters such as the maximum velocity, acceleration, and deceleration capacities can be assumed within reasonable margins. If the traffic controller instantiates a norm, then the subjected vehicle is notified of

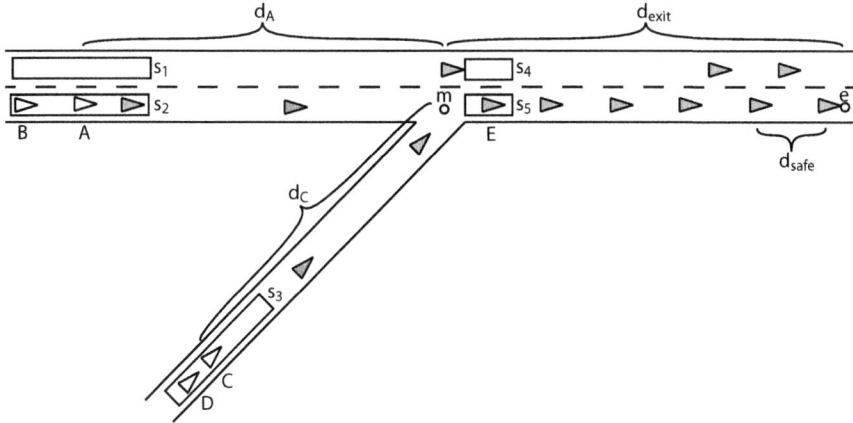

Fig. 2 Scenario with two lanes

the directive that it has to fulfill and the associated sanction that will be imposed if the directive is not followed.

The standard traffic rule in the ramp-merging scenario is that the stream that originates on the main road has priority over the ramp road's stream. However, if the main road is busy, this may lead to large traffic jams on the secondary road. Therefore, the traffic controller uses the traffic data from sensors 1–3 and calculates the optimal velocities of the vehicles in such a way that the traffic streams merge smoothly together at the merge point. More specifically, the traffic controller notifies vehicles passing sensor 1 to stay on the left lane. Sensors 2 and 3 are used by the traffic controller to coordinate the scheduling of vehicles on the merge point. If a vehicle passing sensor 2 has to slow down too much, i.e., to a velocity less than a given threshold, then it receives the directive to move to the left lane.

In order to not overcomplicate the scenario, we decided to simplify some aspects of the traffic controller. The sanction that a vehicle can receive for violating a norm is modeled by either a low or a high fine. A directive that a vehicle can receive is an obligation to be on the left or right lane of the main road at a certain target velocity. For instance, $(right, 10)$ is read as the obligation to be on the right lane at 10 m/s. Given that autonomous vehicles are accurate, we take real numbers for the specification of target velocity rather than multiples of 10 km/h as is common in current traffic controllers. A norm instance consists of a directive that is paired with a sanction. The sets of possible sanctions, directives, and norm instances for this scenario are global throughout the paper and given by:

- $S = \{low, high\}$ are the possible sanctions.
- $O = \{left, right\} \times \mathbb{R}$ are the possible directives.
- $N = O \times S$ are the possible norm instances.

For each simulation step, the new norm instances are created. Recall that norm instances are created based on norm schemes.

3.1 Norm Scheme: Example

For our scenario, the traffic controller instantiates a norm scheme into a vehicle-specific norm in such a way that vehicles cannot arrive on the merge point at the same time if they comply with their norms. This will cause vehicles to slow down considerably on the main road. If they have to slow down too much, then they will be obliged to move to the left lane which is assumed to be more free-flowing. The exact explanation of this norm is provided in Appendix 1.

We also have another norm scheme that obliges vehicles that enter on the left lane of the main road to stay on their lane at a preset maximum velocity v_{MAX}. We shall illustrate the different aspects of a norm scheme according to this scheme's pseudocode that is given in Algorithm 1. We assume that there exists a current set of norm instances N, a function sanction that given a vehicle and fine issues the fine for that vehicle, and a function read that returns the vehicles on a sensor's area that has not been seen before by that sensor. Though not a forced pattern, we do encourage future users of our extension to use the same code structure as in Algorithm 1. Every norm scheme has a specification of when norm instances are created, when they are retracted, and when a sanction should be issued.

Algorithm 1 Pseudocode for the stay-on-lane norm scheme

```
 1: L ← read(s₁)
 2: for all agent ∈ L do
 3:     N ← N ∪ {(agent, ((left, v_MAX), low))}
 4: L ← read(s₄)
 5: for all agent ∈ L do
 6:     if (agent, ((left, v_MAX), low)) ∈ N & agent.v < v_MAX then
 7:         sanction(agent, low)
 8:     N ← N \ {(agent, ((left, v_MAX), low))}
 9: L ← read(s₅)
10: for all agent ∈ L do
11:     if (agent, ((left, v_MAX), low)) ∈ N then
12:         sanction(agent, low)
13:     N ← N \ {(agent, ((left, v_MAX), low))}
```

The code of Algorithm 1 is executed after every simulation tick. We begin with creating new norm instances (lines 1–3). In this case, sensor 1 is read. In Fig. 2, it can be seen that this sensor detects all vehicles that enter the scenario on the left lane of the road. Hence, we give each vehicle on the left lane the directive to stay on the left lane and also obtain a preset maximum velocity to ensure that the flow stays high (line 3). We then continue by checking sensor 4 (lines 4–8). Vehicles with instances of this norm that pass sensor 4 fulfilled their directive to stay on the left lane. However, if their velocity is not the obliged velocity, they receive a low fine (lines 6–7). After passing this sensor, the vehicles are relieved of their directives (line 8). The same holds for sensor 5 (line 9). However, if a vehicle passes sensor 5 then it means that it switched to the right lane. Hence, each vehicle that has received a directive to stay left but passes sensor 5 is fined (line 12).

4 Norm-Aware Vehicles

Our goal is to build autonomous norm-aware vehicles that operate on smart roads where vehicles' behaviors are automatically monitored, evaluated with respect to traffic norms, and possibly sanctioned. As mentioned before, we assume that individual vehicles have their own objectives (e.g., destination, arrival time, travel cost) and are able to deliberate and decide on actions that achieve their objectives. This implies that a vehicle can choose to obey/violate a norm, when this contributes to the achievement of its objectives. In this section, we will discuss norm awareness and how we adopted it in our own driver models.

4.1 Background on Norm Awareness

We have explained norm-based controllers in a multi-agent system as being external entities to the agents, which may oblige/forbid certain behavior and impose sanctions. This abstraction comes from the human way of organizing. It is beneficial to make agents in a multi-agent system norm-aware, especially in simulations where we want the agents to change their behavior due to the norms as humans do. Norm awareness falls under the umbrella of organizational awareness [20]. Organizational awareness is about allowing agents to reason about their role within an organization. However, for traffic simulation purposes, all agents have the same role (i.e., being a driver), hence we will focus on norm awareness only.

Being norm-aware means that an agent can reason about the norms that it receives. The directives that an agent receives are unlikely to match its goals and desires. Otherwise, there would be no need for sanctions. Reasoning about norms hence entails weighing for a course of actions the benefits (reaching goals, fulfilling desires, etc.) and the penalties (expected sanctions from the norm-based controller). In general, the more complex the planning mechanism of the agent is, the harder it is to incorporate reasoning about norms. For an example language for programming norm-aware agents that can deliberate about norms w.r.t. their own goals and plans, we refer the reader to N-2APL [3].

4.2 A Norm-Aware Driver Model

In order to model norm awareness, we deviate from the standard SUMO driver models [14]. There are several reasons for this. First, vehicles in SUMO are goal-directed only in a limited way. For example, the goal of SUMO drivers is to follow a certain route, as opposed to the goal of having a specific location as the destination. Second, SUMO agents are preprogrammed to follow a specific route. They only respond reactively to their environment, instead of deliberating on what action would best suit them. Third, vehicles in SUMO are inherently incapable of deciding to break the rules. For example, they always stop for a red light and they obey to the right of way. This is because of vehicles in SUMO move according to specific car-following and lane-changing rules. The car-following model is not created with norm-aware vehicles in mind. The most commonly used car-following model made by Stefan Krauss is designed purely to create realistic traffic flows in general since in most traffic simulations individual movement on the microscopic level is not interesting [14]. In contrast, we aim at designing futuristic autonomous cars with a more fine-grained sense of control and, most importantly, the ability to violate norms. In our work, we model drivers with different possible actions, a belief state, and personal preferences. The available actions of a vehicle depend on the scenario. The belief state of a vehicle consists of:

- A description of its runtime variables which contain its velocity, position (given by a lane and distance from that lane's start), and current norm instances.
- An expected arrival time (at the goal location) function that reflects, for instance, the GPS planning tools that vehicles have available. This function is equal for all vehicles, but can be parameterized in the future in order to make more optimistic/pessimistic drivers.
- A local effect function that returns the next expected runtime variable configuration, given the runtime variables of a vehicle and an action. For instance, if the current velocity is 20 m/s and a vehicle accelerates by 5 as an action, then it expects for the next simulation tick to be at 25 m/s if this is possible within its acceleration capabilities. This function is also equal for all vehicles.
- A directive distance function that returns a positive expectancy of whether the vehicle can fulfill the directive in time before it is sanctioned, given runtime variables and a directive. The precise definition of this function depends on the possible directives and driver specifics in an application. However, a high distance should mean that it is likely that the sanction will be incurred in the future, whereas a distance of zero should indicate that the current state fulfills the directive. This function is also equal for all vehicles.

The personal preferences of a vehicle are given by its personal profile. This profile consists of:

- A maximum desirable velocity of the vehicle.
- A sanction grading function that returns how bad a sanction is to the vehicle. The higher the number, the worse the sanction. This can be used to model how affluent or greedy a vehicle is.
- A arrival time grading function that returns how good or bad an arrival time is. This encodes the desired arrival time of the vehicle. Anything before the desired time

Fig. 3 Sense–reason–act cycle of an agent

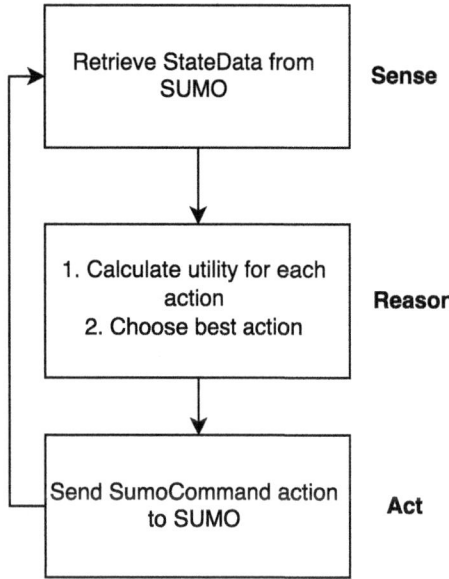

should be evaluated to zero or less, and everything after it should monotonically increase in their evaluation.

Our vehicles are assumed to use a sense–reason–act cycle. In our extension, this cycle is performed at each simulation tick of SUMO (Fig. 3). In practice, it is not required that the vehicles run synchronously with the traffic controllers and/or other vehicles. In practice, a vehicle perceives its road environment by its onboard sensors. In our extension, this information is obtained from SUMO, which gives a vehicle its local information. A vehicle also receives newly instantiated or retracted directives that apply to itself. This information updates the vehicle's belief state (its runtime variables). Using its knowledge of the road network, the vehicle estimates the utility of each of its actions by balancing between the value of achieving its objectives and possible sanctions that will occur if it performs the action. The arrival time and expected sanctions are used to calculate the utility for each action.

The action with the highest utility is chosen by the agent's action selection function, using a priority-based tie-break mechanism for ties. The tie-break mechanism is used when two or more actions have the highest utility. If such a situation occurs, then the action with the highest priority is chosen. In our implementation, the less an action changes the agent state, the higher its priority is. For example, in a tie-break situation, doing nothing is preferred to increasing velocity to a small amount, which in turn is preferred to increasing velocity to a larger amount, which in turn is preferred to changing lane. We chose for this 'least impactful action' tie-break ordering since we believe this to be in line with human behavior. However, we stress that this ordering is not essential to our framework and can be replaced by arbitrary tie-break orderings. Finally, the simulator executes this action.

Recall that S is the set of possible sanctions, O the set of possible directives, and N the set of possible norm instances. The relevant components of an agent are modeled as follows:

- A is the set of actions to choose from.
- $\langle v, l, d, n \rangle$ is a specification of the runtime variables of a vehicle, where v is the current velocity, l is the current lane, d is the distance from the starting point of that lane, and $n \subseteq N$ are norm instances.
- B is the set of all possible runtime variables configurations.
- $f : B \mapsto N$ is the expected arrival time function.
- $e : B \times A \mapsto B$ is the local action effect function.
- $\delta : B \times O \mapsto \mathbb{R}$ is the directive distance function.
- $\langle v_{max}, g_s, g_t \rangle$ is a specification of a vehicle's personal profile, where v_{max} is the maximum velocity, $g_s : S \mapsto \mathbb{R}$ is the sanction grading function, and $g_t : N \mapsto \mathbb{R}$ is the arrival time grading function.
- P is the set of all possible personal profiles.
- $u : B \times P \times A \mapsto \mathbb{R}$ is the utility function, given by:

$$u(b, p, a) = g_t(f(b')) + \sum_{(o,s) \in n} (\delta(b', o) \cdot g_s(s)),$$

where $b = \langle v, l, d, n \rangle$, $p = \langle v_{max}, g_s, g_t \rangle$, and $b' = e(b, a)$.
- $\alpha : B \times P \to A$ is the action selection function given by:

$$\alpha(b, p) = \max_{a \in A} u(b, p, a)$$

5 Application of Norm-Aware Driver Models

In this section, we shall go into detail how the norm-aware driver models are specified for our scenario. We give a specification of the scenario's specific components and discuss utility calculations.

5.1 Vehicle Driver Specification

A vehicle reasons about all its actions when it deliberates for a next action. Among all actions in our scenario are de-/accelerating action. We have simplified this by discretizing the possible de-/acceleration values. The other possible actions available to a vehicle are switching a lane to the left or right. More specifically, the set of actions A that a vehicle can decide to perform are:

$$A = \{a_x \mid x \in \{0, 0.1, -0.1, 1, -1, 5, -5, 10, -10, 20, -20, 50, -50\}\} \cup \{l_{left}, l_{right}\},$$

where a_x is read as adding x to the current velocity (i.e., de-/accelerating) and l_{left}/l_{right} is read as moving a lane to the left or to the right.

We also specify how the directive distance measure is calculated given the possible directives in our scenario. In our scenario, the factors that a vehicle considers are the time it takes to fulfill the directive, the current time, and the expected time that the sanction will be issued if the directive is not followed. We have implemented vehicles in such a way that they expect that the traffic controller check whether a directive is followed somewhere between the current time t and the current expected exit time t_{exit} for the vehicle. The minimal amount of time needed to adhere to the norm is denoted δ_t. For instance, if a vehicle at time step t is being instructed to drive 25 m/s and can accelerate to this speed in minimally three time units, then $\delta_t = 3$. If we need zero steps to adhere to the norm, the distance is zero. Otherwise, the distance proportionally moves to 1 given the current time. If the traffic controller will check directive fulfillment before the driver can achieve compliance (i.e., $t_{exit} - t < \delta_t$), then δ should be 1. Hence, the directive distance measure is given by:

$$\delta(b, (l_o, v_o)) = \frac{\min(\delta_t, t_{exit} - t)}{(t_{exit} - t)},$$

where t is the current time, t_{exit} is the expected exit time, and δ_t is the minimal number of steps needed for compliance of (l_o, v_o) given the current runtime variables b.

Note that this means that the drivers expect the traffic controller to issue a sanction if a directive is not followed between now and $t_{exit} - t$ time units later. This distance measure can be modified easily in our framework.

5.2 Utility Calculations: Example

To illustrate this notion, suppose we have two drivers, a poor one and an affluent one in an identical situation. They currently drive 20 m/s, their maximum speed is 30 m/s, they can accelerate or decelerate with 10 m/s and their travel distance is 1080 m. Both are in a hurry, so their g_t is defined as $g_t(time) = \frac{bestTime}{time}$.

Here, $bestTime$ is defined by the minimal travel time, i.e., the time it would take the drivers to travel the distance if they could go their maximum speed all the time. In this case, $bestTime = 1080/30 = 36$. However, the road the drivers travel on has a speed norm, with the maximum speed being 10 m/s. Not complying with this norm gives a high fine. The poor agent cannot afford this fine, so it has $g_s(high) = -20$. The affluent driver can easily afford this fine, so it has $g_s(high) = -0.2$. Suppose for this example that drivers can only take the actions a_0, a_{10} and a_{-10}.

In Table 1, we see the utilities for each of these actions for both drivers. Here, we see that the highest rewarded action for the poor driver is the one that obeys the norm since it cannot afford the fine, while the affluent driver is in a position to violate the

Table 1 Example of the deliberation of a poor and affluent agent

	a_0	a_{10}	a_{-10}
Speed after action	20 m/s	30 m/s	10 m/s
Norm speed	10 m/s	10 m/s	10 m/s
Travel time remaining	54 s	36 s	108 s
g_t	0.67	1.00	0.33
Steps needed to oblige the norm	1	2	0
δ	0.02	0.06	0.00
Utility poor agent	0.30	−0.11	0.33
Utility rich agent	0.66	0.99	0.33

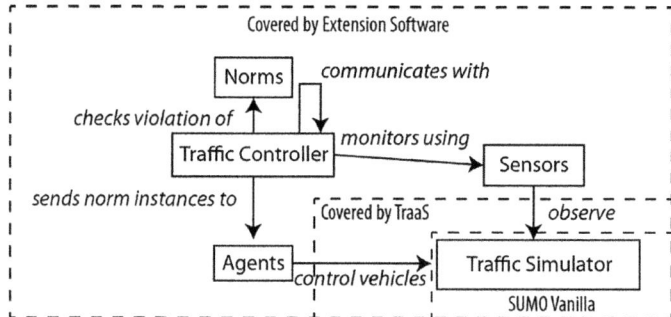

Fig. 4 Overview of the presented extension

speed norm since it can afford the fine. In fact, the affluent driver will increase its speed since it then maximizes its time grade and thereby its utility.

6 Implementation Approach

The structure of the presented norm-based traffic controller for SUMO is depicted in Fig. 4. As stated before, we use the native SUMO application as the environment. However, we do not use SUMO's driver models. Instead, we communicate commands from our own agent model to the vehicles. We have also implemented our own sensor business logic. These sensors are connected to lane area detectors as implemented in SUMO. The communication between the extension software and SUMO is provided by the TraaS library. We use and provide a new version of TraaS. In addition to improved performances, the new version has some extra functionalities that are needed for our extension.

The software in our extension is composed mainly of the agent models and the traffic controller. The software executes in lockstep with SUMO; i.e., each simulation

Fig. 5 Model-View-Controller structure

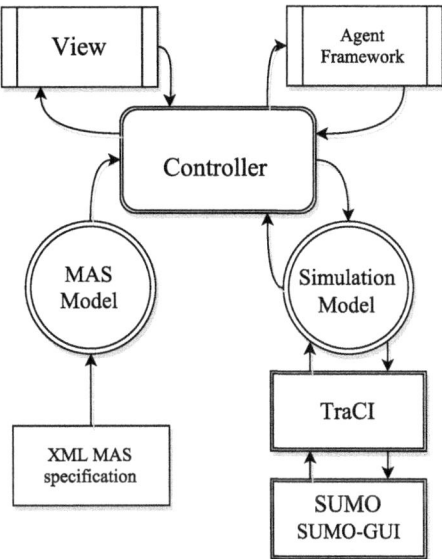

tick in SUMO is also a tick in the extension. We chose to provide our work as an extension to SUMO, rather than modifying the source code of SUMO directly.

There are various pragmatic reasons to propose a SUMO extension rather than altering its code. For example, we can now support multiple versions of SUMO, starting from SUMO 0.20 and upwards. This also makes our extension more accessible to users, since it eliminates the need for users to recompile SUMO before they can use the framework. The framework is developed in Java since most norm-based autonomous agent frameworks are written in Java and Java programs are easy to use on a variety of platforms. We have designed the framework using the Model-View-Controller pattern, as can be seen in Fig. 5. A multi-agent system can be specified in XML and is converted to a MAS model. We furthermore use TraaS to retrieve the simulation state of SUMO. These models can be manipulated by the controller software (note that the controller is a control component in the software engineering sense; it is not the same as the conceptual traffic controller). The agent framework simulates our driver models. The view of the system allows the user to see the state of the MAS side of the simulation. This decomposition allows other researchers to easily create their own front-end by changing the view implementation. It is also possible to create a new data format for scenarios by changing the data model implementation, or to change the simulation package by providing a different simulation model implementation. Each part can be changed without the need to change other parts of the framework.

In Fig. 6 an UML overview of the traffic controller structure is shown. As mentioned previously, traffic controllers observe by using sensors. These sensors are an extension of a physical object, occupying space within the environment on a certain road, lane, and distance and have a certain length. The traffic controller consists of

Fig. 6 UML overview of
traffic controllers

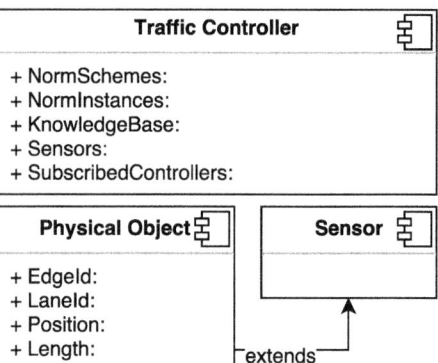

five attributes. First, NormSchemes, which is a set of traffic regulations to be effectuated by the traffic control system. Second, NormInstances, a set containing all norms instantiated by the controller. In effect, this means a specific regulation is sent to a vehicle. Third is the set Sensors, which is a set of sensors at its disposal to sense the vehicles driving in the environment. Fourth, a traffic controller is subscribed to a set of other traffic control systems denoted by the set SubscribedControllers. Finally, KnowledgeBase is a set containing data about the vehicles in the simulation.

At regular intervals, vehicle data is collected from both the sensors and other traffic controllers and is used to update the traffic controller's knowledge base. It uses its knowledge base in three ways. The first usage is to decide if the traffic regulation should be applied and instantiates a specific regulation for a vehicle. Secondly, a traffic controller employs its knowledge base to determine if the regulation is violated in order to impose sanctions. Finally, the knowledge base is used to determine whether the deadline of an instantiated regulation is met. If so, the regulation is retracted and the relevant vehicle is notified.

The communication of the regulations to vehicles as well as the communication of sensor readings between traffic controllers is assumed to be done wireless. The actual communication protocol is unspecified.

An UML overview of the agent structure is shown in Fig. 7. Agents are an extension of physical objects since they occupy a certain space and have a certain position in the world. A Basic Agent is a prototype AgentProfile. The agent profile is specified by the SanctionGrade and ArrivalTimeGrade functions which are the sanction and arrival time appraisal functions discussed in Sect. 4.2. In the doAction function, the directive distance measure is used to calculate the utility for each individual action. In this manner, the agent profile of the agent together with the current state decides what action the agent will choose toward achieving its goal. These design choices were made to model the agents in such a way that new agent profiles can easily be created to model norm-aware human or driverless autonomous vehicles, while still keeping the agents as simple as possible. With each deliberation cycle, agents can pick the action most suited to their goals and the current state, one of the actions listed by the AgentAction enumeration in Fig. 7.

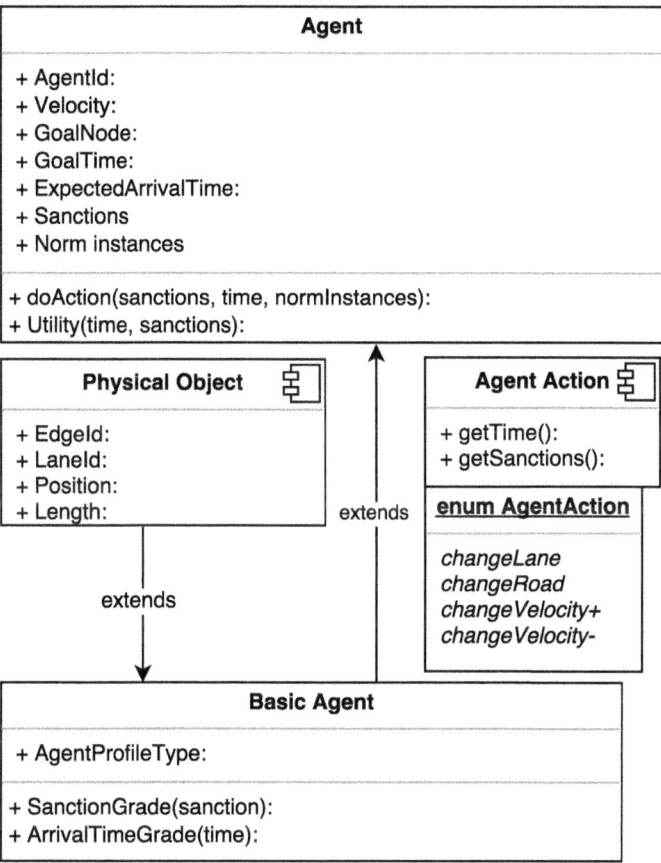

Fig. 7 UML overview of the agent structure

7 Experimental Evaluation

We tested the performance of our normative agent-based traffic approach using four experiments. In these experiments, variations of the ramp-merging scenario, as explained in Sect. 3, are used. The first experiment considers a ramp-merging scenario where the main road consists of a single lane, while on the second, third, and fourth experiment, the main road has two lanes. In the second experiment, the second lane is accessible for all drivers, but in the third experiment the second lane is marked as an emergency-only lane. Finally, in the fourth experiment, the ramp-merging scenario is used twice in succession in order to demonstrate the use of communication and coordination between decentralized traffic controllers.

The experiments were set up as follows. Each experiment is run for a length of one hour (3600 ticks). The spawn rate shown in the tables of the experiments is defined as the chance of a vehicle spawning every tick. If there is not enough room to spawn a vehicle at a certain time, then SUMO puts the vehicle on hold and spawns it at the earliest possible time when space is available. The maximum speed at the merge point, v_{max}, was set to 80 km/h. Furthermore, the four experiments consist of comparing two scenarios, both ran for one hundred times. The values displayed in the tables are the averages over hundred runs. The throughput is defined as the number of vehicles leaving the simulation every tick. Average speed is the average speed over all runs in m/s, and finally the average gap is the average distance between two cars in meters. Also, we define for each experiment the maximum (expected) throughput. This is the expected throughput if each vehicle could keep driving its maximum speed throughout the scenario and can be calculated by the following formula: $throughput_{max} = 60p$, where p is the probability of a car entering the simulation on that tick.

7.1 Experiment 1: SUMO and TrafficMAS

The first experiment illustrates the behavior difference between the default SUMO vehicles and the norm-aware driver models implemented in the TrafficMAS extension. A single traffic controller observes the vehicles in the simulation and communicates tailored norm instances to each vehicle. In this experiment, a norm instance is simplified to just a target velocity since there is no choice of lanes on the main road. The expected result is that norms result in a higher average velocity and a better throughput of vehicles since traffic jams will be prevented. In this scenario, a classic ramp-merging situation is implemented, where both the main road and ramp consist of a single lane. The spawn rate of the vehicle input stream will be slightly higher on the main road to resemble a realistic traffic situation. In the TrafficMAS scenario, three sensors are placed on the road: one on the main road, one on the ramp, and a control sensor on the output road. The traffic controller instantiates norms, removes norms, or applies sanctions when the vehicles are detected by the sensors. In the SUMO scenario, the main road has priority over the ramp road, comparable to real-life merging situations.

As is clear from the results in Table 2, there is an increase in both throughput, average speed, and the average amount of space between the vehicles. This is the case since in the SUMO scenario a traffic jam instantly forms on the ramp, because of the relatively high density of cars on the main road (Fig. 8). These results confirm our expectation of coordination by a norm-based traffic controller improving on classic ramp-merging scenarios. Note that the throughput % value exceeds a hundred percent, this is possible because the spawn rate is probability based and thus can exceed the maximum expected throughput.

Table 2 Results for the first scenario

	SUMO agents	Norm-aware agents
Main road spawn rate	20%	20%
Ramp spawn rate	15%	15%
Throughput	16.16	21.01
Max throughput	21	21
Throughput %	76.95%	100.05%
Average speed	3	20.97
Average gap	13.81	101.82

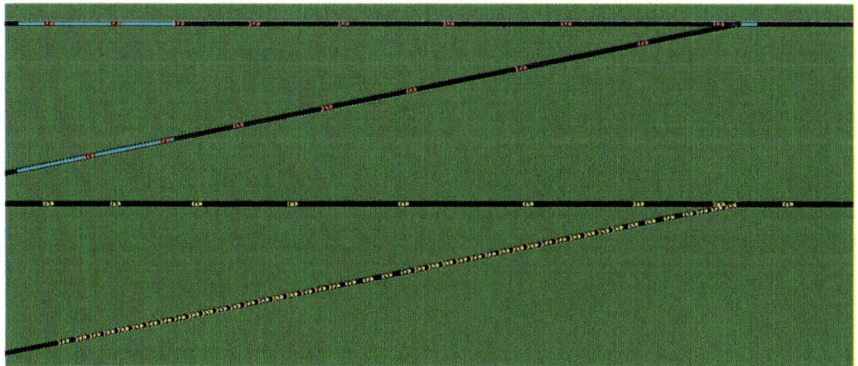

Fig. 8 Screenshot depicting the difference in performance in experiment 1. The top scenario uses our framework and merge norm. The bottom scenario uses the default SUMO driver models

7.2 Experiment 2: Simple Norms and Advanced Norms

The goal of the second experiment is to compare traffic controllers using simple and advanced norms. The traffic controller in the SingleNorm scenario observes and controls the same norm as in experiment 1. In the AdvancedNorm scenario, the traffic controller can also issue directives for the vehicles to change lanes in order to relieve the rightmost lane traffic and prevent congestion. The lane change directive will be given to a vehicle when its calculated velocity on the merge point is below a certain threshold. For this experiment, the threshold was set to $1/2v_{max}$. Our expectation is that in this multi-lane scenario, the traffic controller with the advanced norm can successfully cope with a higher input stream of vehicles.

The setup for the SingleNorm scenario is a copy of the TrafficMAS scenario in experiment 1, except that in this case the main road has two lanes instead of one, and moreover, the input stream of vehicles of both roads are increased. The AdvancedNorm scenario implements extra sensors on the second lane, but is exactly the same in every other aspect.

Table 3 The results for the second scenario

	SimpleNorm	AdvancedNorm
Main road spawn rate	30%	30%
Ramp spawn rate	20%	20%
Throughput	20.38	29.91
Max throughput	30	30
Max throughput %	67.93%	99.70%
Average speed	3.31	14.91
Average gap	14.2	61.49

As can be observed from the results in Table 3, the simple norm cannot cope properly with the increased spawn rate of vehicles in this scenario. The average speed has diminished severely, as well as the average gap between vehicles. This means congestion is abundant in the SimpleNorm scenario. However, the AdvancedNorm seems to cope very well with the increased input stream of vehicles. In this scenario, the throughput approximates the maximum expected throughput by a factor 0.3%, which indicates that the vehicles move throughout the simulation without much congestion.

7.3 Experiment 3: Sanction Severity

The third experiment illustrates that drivers are able to reason about norms. Experiment 1 has shown that drivers are norm-aware. However, TrafficMAS agents also have the capabilities to violate norm if these violations do not have significant impact on them. In this experiment, the leftmost lane is an emergency lane, reserved for certain traffic in order to help with accidents and other emergencies. Therefore regular drivers will get sanctioned if caught driving on this lane. Since this lane remains mostly empty, this is a viable option for drivers who greatly value a faster arrival time and are in a financial position which makes them willing to accept a fine. We expect that the more affluent drivers will choose to accept sanctions in order to improve their arrival time, resulting in distinct behavior between the two groups of drivers.

This experiment is set up in the same way as experiment 2, except that the leftmost lane is reserved for emergencies and the spawn rates are lowered. In the Poor Drivers scenario, the input stream consists of drivers who are impatient, but in a substandard financial position. The Affluent Drivers scenario spawns drivers who care about sanctions, but are willing to accept fines if by doing so they can arrive earlier to their destinations.

The results of this experiment are listed in Table 4. With this experiment, the differences in throughput, average speed, and average gap are much smaller, and

Table 4 The results for the third scenario

	Poor drivers	Affluent drivers
Main road spawn rate	20%	20%
Ramp spawn rate	15%	15%
Throughput	20.48	20.88
Average speed	12.95	14.42
Average gap	48.37	69.29
Sanctions	0	133.12

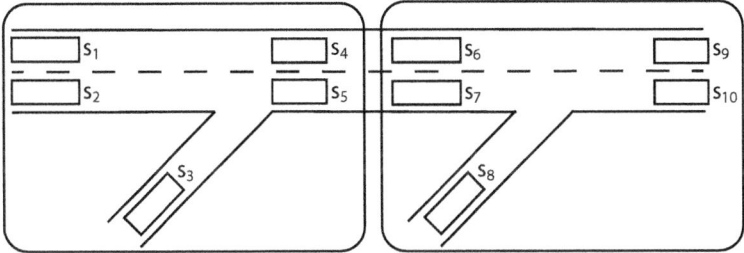

Fig. 9 Distributed traffic control setting. Rounded boxes indicate local traffic controllers. The left controller is connected to sensors 1–5 and the right controller to sensors 6–10

not significant enough to lead to any conclusions about improvement. However, a significant distinction in the number of sanctions can be seen. This indicates a difference in behavior between the groups of drivers. On average about 133 affluent drivers decide to drive on the emergency lane in an hour of simulation. This shows a clear difference in behavior from the poor drivers, who never decide to change lanes.

7.4 Experiment 4: Communication in Distributed Traffic Control Systems

The final variation of the merging scenario that we consider demonstrates the decentralized version of our framework. Specifically we demonstrate the ability of one traffic controller to share data about traffic with another traffic controller so that the receiver can adjust its norms. In this variation there are two merge points in sequence (Fig. 9). Each of the merge points is controlled by a local traffic controller as in the previous scenarios. For this, they have their own local sensors.

When running this scenario without communicating traffic controllers, we observed that the traffic streams tend to flow like the top situation in Fig. 10. Traffic that arrives on the left lane of the main road keeps that lane, as it is faster than switching to the right lane. The ramp traffic streams merge in on the right lane of the main road. After the merge scenario, the vehicles can freely move from left to right and back.

Fig. 10 Top: Traffic streams (arrows) without coordination. Bottom: traffic streams with coordination

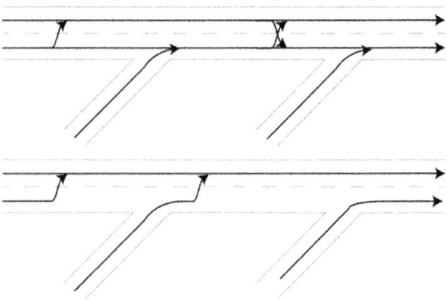

However, if the stream of vehicles in the second ramp is too dense, then congestion occurs at the second merge point. The problem is that the second merge point has to process too many vehicles.

A solution might be to redirect all traffic observed by the left traffic controller to the left lane of the main road when high-density streams are observed at the ramp road of the right traffic controller (the second ramp road). This way all traffic on the second ramp road can continue through on the right lane of the main road without being obstructed by oncoming traffic. However, the left traffic controller can only sense the traffic situation using its local sensors. Therefore the right traffic controller needs to inform the left traffic controller about the traffic density on the second ramp.

This coordination is realized as follows. The left traffic controller subscribes to traffic density observed by sensor s_8 of the right traffic controller. If the left traffic controller detects a high traffic density on the second ramp, it will issue new norm instances that obliges vehicles to move to the left lane. As a result the input traffic streams of the right traffic controller should be easily manageable as the vehicles on the main road are obliged to stay left such that the vehicles on the second ramp roads can move on the right lane of the main road. The resulting traffic streams should resemble the streams in the bottom depiction of Fig. 10. We expect that the coordinating traffic controller performs better in terms of throughput, average speed, and average gap, since less congestion should occur at the second merge point.

The results of experiment four are listed in Table 5. A small increase in the throughput and a larger increase the average speed and gap in the coordinated traffic controllers scenario compared with uncoordinated traffic controllers scenario can be observed. Thus, giving vehicles on the main road the obligation to change to the left lane quickly after the first merging point appears to prevent the delays as observed in the original scenario. These preliminary results support our hypothesis that observation sharing and communication between traffic control systems can be effective for traffic regulation.

Table 5 The results for experiment four

	No coordination	Coordination
Main road spawn rate	25%	25%
Ramp #1 spawn rate	15%	15%
Ramp #2 spawn rate	35%	35%
Throughput	43.81	44.74
Max throughput	45	45
Max throughput %	97.36%	99.36%
Average speed	11.32	14.60
Average gap	62.47	66.03

8 Contributions and Comparison to Other Work

The results of our reported experiments were positive in the sense that the advanced version of our framework performed better than the SUMO baseline/simpler versions of our framework. However, as noted previously, some scenarios were not completely realistic. The merge scenario did not have an acceleration lane aligned with the main road. Moreover, in the final experiment, the traffic density of the second merge lane was higher than one would expect in the real life. However, the aim of our experiments was not to show simulate calibrated realistic scenarios. Our goal was rather to show that our extension is an enabling technology for the specification and testing of norm-based traffic control, which can be extended into a more complex and feature-rich framework.

The contributions of our work are as follows. First of all, we have created a lightweight framework for autonomous norm-aware vehicles and norm-based traffic controller on top of SUMO. This framework is easy to extend with different types of driver profiles. It also allows for the easy usage of a different simulation package. Secondly, we have created the possibility to conduct traffic experiments and measuring the impact of a norm-based traffic controller. Finally, we have improved the TraaS performance, yielding a performance increase of up to four times over the original TraaS library in certain scenarios.

Our approach has some similarity with Baines et al. [6] since they employ autonomous agents and use governing institutions to influence agents to have desirable behavior. However, Baines et al. concentrate on agents' internal architecture, situational awareness, and the communication between agents. The project is set up with realistic maps imported from the Open Street Map foundation and used real-world data from a highway in the UK, the M25.

While our framework is related to the work done by Baines et al., the aim of our research is different. Our driver model is deliberately kept simple in order to focus on the interaction between traffic controllers on the one hand, and between agents and traffic controller on the other hand. Furthermore, our framework is not developed in order to simulate the existing real-world scenario. Finally, our framework supports decentralized traffic controllers while Baines et al. focus on a single, all-knowing, institution.

Another comparable line of research has been done by Balke et al. [7]. In this extended abstract, Balke et al. discuss the difference between off-line and on-line reasoning of institutions (similar to norm-based traffic controllers) governing open multi-agent systems. They state that most research up to that point had been focused on the off-line reasoning of institutions, which can be used to research the static properties of institutions. The on-line reasoning of institutions concerns the monitoring and controlling of agents, observing if norms are being violated and informing agents if this is the case. In this implementation, there is a single institution with the title "The Governor" with which agents can communicate and receive information regarding possible consequences of their actions.

Our approach is most related to the on-line reasoning as described by Balke et al. However, communication between the agents and the institution is handled in a different way. With our framework, the information provided by agents to the traffic controller is acquired via sensors. This is a more realistic representation of traffic situations, since it is often beneficial for the agent to not disclose any information about itself. Furthermore, in TrafficMAS, multiple traffic controllers are present, creating a more robust and better-controlled system through communication and coordination between these institutions.

9 Future Work

Our framework could be extended in a number of ways. One can add support for contrary to duty norms. Contrary to duty norms consist of a hierarchy of norms. An agent should comply with all hierarchies, but if it does not comply with the first norm level (and thus incurring a sanction), it should at least comply to the second norm level, or get an even higher sanction. An example is the norm "You shouldn't break the speed limit, but if you do, you should drive on the leftmost lane."

Secondly, we want to implement a full-fledged decentralized norm-based traffic controller. Currently TrafficMAS supports only decentralization of monitoring in the sense that sensor data can be shared. However, we want to also steer toward a system where norm instances, meta-analyses of sensor data, and sanction commands can be sent between traffic controllers.

Finally, a graphical user interface (GUI) could be added to (i) allow for easy creation of scenarios and (ii) allow for on the fly monitoring and editing of norms. For example, with a GUI one could investigate how the vehicle behavior changes when one changes the sanction severity of a norm. Because of our Model-View-Controller structure of the framework, this can be implemented by only altering the View part of our extension.

10 Conclusion

Our goal was to create a traffic simulation Multi-Agent System where vehicles should generally follow traffic regulations, yet are able to ignore these regulations in certain circumstances without implementing hard constraints on the agents themselves. We used norm-based traffic controllers since they can properly deal with these kinds of situations.

Our work is an extension to SUMO. It features a norm-based traffic controller which monitors and possibly sanctions the vehicles. We assume deliberative proactive drivers that make autonomous decisions according to their goal and received sanctions. The extension features (i) driver profiles which model different types of behavior, (ii) traffic controllers and norms to control vehicles, and (iii) an easy way to add new driver profiles, traffic controllers and norms. This plug and play extension to SUMO can serve as a testing suite for all experiments concerning norm-based traffic control.

We showed in our experiments that the performance of normative systems is better than the default behavior in a ramp-merge scenario. Furthermore, we presented that complex norms allow for finer grained steering of behavior in complicated scenarios. Moreover, we illustrated the autonomy of drivers, by demonstrating a difference in behavior between poor and affluent drivers. Finally, we demonstrated the ability of traffic controllers to coordinate their activities, yielding better results in certain scenarios.

Appendix 1: Merge Norm Scheme

For the merge norm scheme, we use the same pseudocode structure (Algorithm 2) as for the stay-on-lane norm scheme. As with the other norm scheme we begin with the instance of the norm (lines 0–8).7 Initially we read sensors 1 and 3 and merge the readings using the algorithm of Wang et al. [21] (line 1). The result is an ordered list of agents, which, if they continue as they are, will arrive at the merge point in the same order. We maintain a global variable t_{free} that indicates the next moment in time that the merge point is free. With $optimalVelocity$ we calculate the optimal speed for an agent s.t. it will arrive at t_{free} on the merge point plus some safe margin, or later if the agent cannot make it in time physically (line 3). If the agent is at the right lane of the main road and the optimal velocity is below a predefined threshold, then it is obliged to move to the left lane (line 5), otherwise it is obliged to adapt its velocity to the optimal velocity and pass the merge point on the right lane (line 7). An agent is sanctioned if it is not passing the merge point on the correct lane (lines 11–12, and 15–16). Otherwise, an agent can also be sanctioned if it did not achieve its predetermined velocity (lines 19–20).

Algorithm 2 Pseudocode for the merge norm scheme

```
1:  L ← merge(read(s₁), read(s₃))
2:  for all agent ∈ L do
3:      s ← optimalVelocity(agent, t_free)
4:      if agent.lane = right&s < v_threshold then
5:          N ← N ∪ {(agent, ((left, v_MAX), high))}
6:      else
7:          N ← N ∪ {(agent, ((right, s), high))}
8:  L ← read(s₄)
9:  for all agent ∈ L do
10:     N ← N\{(agent, ((left, v_MAX), high))}
11:     if (agent, (right, s), low)) ∈ N then
12:         sanction(agent, high)
13: L ← read(s₅)
14: for all agent ∈ L do
15:     if (agent, (left, v_MAX, high)) ∈ N then
16:         sanction(agent, high)
17:         N ← N\{(agent, ((left, v_MAX), high))}
18:     else if (agent, (right, s), high)) ∈ N then
19:         if s ≠ agent.v then
20:             sanction(agent, high)
21:         N ← N\{(agent, ((right, s), high))}
```

Appendix 2: Using Our Code

Our framework is open-source and available on-line on Github at https://github.com/baumfalk/TrafficMAS. It can be compiled from source, or it can be downloaded as a binary version.

How to Run It

Our framework can be run as follows. Assuming you use the binary JAR file, a scenario can be run with the following command:

```
java -jar TrafficMAS.jar ./scen/ scenario.mas.xml
path/to/sumo scenario.sumocfg [seed].
```

In this command `scen` is the directory the scenario is located in, `scenario.mas.xml` is the main configuration file for the scenario and `path/to/sumo` denotes the SUMO executable to use. The SUMO-GUI program can also be used. The parameter `scenario.sumocfg` denotes the SUMO configuration file used by the scenario. Finally, the parameter `seed` is used to prepare the random number generator, which is used to spawn vehicles in a probabilistic fashion. If no seed is provided, a random one is generated by the system.

How to Create Your Own Scenario

Our framework also allows for the creation of your own scenarios. A TrafficMAS scenario consists of several XML files:

- a global configuration file, containing the paths to the other XML files, as well as the simulation duration.
- a configuration file specifying which norms are used. In this file, the norms are also parameterized with scenario-specific information, such as road names.
- a configuration file which describes the norm-based traffic controllers. The file is used to define which controllers there are, which sensors they have access to and to which other controllers they are subscribed.
- a configuration file containing the vehicle profile distributions. This file contains the distributions of the various driver profiles and the traffic density of the different roads.
- various SUMO XML files: the XML file containing the nodes, the edges, the sensors and the routes.

References

1. Abdelkader G (2003) Requirements for achieving software agents autonomy and defining their responsibility. In: Proceedings of the autonomy workshop at AAMAS, vol 236
2. Ackerman E (2015) Tesla working towards 90 percent autonomous car within three years. http://spectrum.ieee.org/automaton/robotics/artificial-intelligence/tesla-working-towards-90-autonomous-car-within-three-years. Accessed 29 June 2015
3. Alechina N, Dastani M, Logan B (2012) Programming norm-aware agents. In: Proceedings of the 11th international conference on autonomous agents and multiagent systems-volume 2, International foundation for autonomous agents and multiagent systems, pp 1057–1064
4. Alechina N, Dastani M, Logan B (2013) Reasoning about normative update. In: Proceedings of the twenty-third international joint conference on artificial intelligence. AAAI press, pp 20–26
5. Baines V, Padget J (2014) On the benefit of collective norms for autonomous vehicles. In: Proceedings of 8th international workshop on agents in traffic and transportation
6. Baines V, Padget J (2015) A situational awareness approach to intelligent vehicle agents. In: Modeling mobility with open data. Springer, Berlin, pp 77–103
7. Balke T, De Vos M, Padget J, Traskas D (2011) On-line reasoning for institutionally-situated bdi agents. In: The 10th international conference on autonomous agents and multiagent systems-volume 3, International foundation for autonomous agents and multiagent systems, pp 1109–1110
8. Baskar LD, De Schutter B, Hellendoorn J, Papp Z (2011) Traffic control and intelligent vehicle highway systems: a survey. IET Intell Transp Syst 5(1):38–52
9. Boella G, Van Der Torre L, Verhagen H (2006) Introduction to normative multiagent systems. Comput Math Organ Theory 12(2–3):71–79
10. Dastani M, Grossi D, Meyer J-JC, Tinnemeier N (2009) Normative multi-agent programs and their logics. In: Knowledge representation for agents and multi-agent systems. Springer, Berlin, pp 16–31
11. Hübner JF, Boissier O, Bordini RH (2011) A normative programming language for multi-agent organisations. Ann Math Artif Intell 62(1–2):27–53

12. Kavathekar P, Chen Y (2011) Vehicle platooning: a brief survey and categorization. In: ASME 2011 international design engineering technical conferences and computers and information in engineering conference. American society of mechanical engineers, pp 829–845
13. Krajzewicz D, Erdmann J, Behrisch M, Bieker L (2012) Recent development and applications of sumo–simulation of urban mobility. Int J Adv Syst Meas 5(3–4)
14. Krauss S, Wagner P, Gawron C (1997) Metastable states in a microscopic model of traffic flow. Phys Rev E 55(5):5597
15. Markoff J (2015) Google cars drive themselves, in traffic. http://www.nytimes.com/2010/10/10/science/10google.html. Accessed 29 June 2015
16. Meneguzzi F, Vasconcelos W, Oren N, Luck M (2012) Nu-BDI: Norm-aware BDI agents. In: Proceedings of the 10th European workshop on multi-agent systems. Dublin, Ireland
17. Testerink B, Dastani M, Meyer J-J (2014) Norm monitoring through observation sharing. In: Proceedings of the European conference on social intelligence, pp 291–304
18. Testerink B, Dastani M, Meyer J-J (2014) Norms in distributed organizations. In: Coordination, organizations, institutions, and norms in agent systems IX. Springer, Berlin, pp 120–135
19. Tinnemeier NA, Dastani M, Meyer J-J, Torre L (2009) Programming normative artifacts with declarative obligations and prohibitions. In: Web intelligence and intelligent agent technologies. WI-IAT'09. IEEE/WIC/ACM international joint conferences on 2009, vol 2. IET, pp 145–152
20. van Riemsdijk MB, Hindriks K, Jonker C (2009) Programming organization-aware agents. In: Engineering societies in the agents world X. Springer, Berlin, pp 98–112
21. Wang Z, Kulik L, Ramamohanarao K (2007) Proactive traffic merging strategies for sensor-enabled cars. In: Proceedings of the fourth ACM international workshop on vehicular ad hoc networks. ACM, pp 39–48

Public Transport, Logistics and Rail Traffic Extensions in SUMO

Andreas Kendziorra and Melanie Weber

Abstract Disasters and major events affect the efficiency of passenger and freight transportation. The project VABENE++ considers the question what happens when the transport network is disturbed due such an event. Within this project, the German Aerospace Centre developed a decision support tool that provides consolidated information and operation recommendations regarding individual motor car traffic based on traffic simulation performed by SUMO. Recently, it was aimed to realise simulation scenarios in which incidents impairing the transport network occur, and that focus on multimodal transportation systems. To enable this, the implementation of public transport in SUMO was extended and a new logistic concept was implemented. Furthermore, rail signals were implemented to enable more realistic rail-bound traffic, as this mode of transport is very significant for both public transport and logistics. The present paper presents these extensions in detail and outlines its usefulness in examples.

Keywords SUMO · Logistic · Transport · Public transport · Railway

1 Introduction

Transport and traffic is one of the identified critical infrastructures in Germany [1]. A disturbance of the traffic network can have serious influences on passenger transportation, freight transportation and the supply of necessary goods and services. These disturbances can be manifold. They may be plannable, like major events, or unpredictable, like accidents, natural disasters (e.g. floods, earthquakes) or from human malevolence, such as acts of sabotage, terrorism or war. To support decision makers, like authorities and emergency forces, in such critical situation, the decision support system *EmerT* (Emergency mobility of rescue forces and regular Traffic) was devel-

A. Kendziorra (✉) · M. Weber
German Aerospace Center, Rutherfordstraße 2, 12489 Berlin, Germany
e-mail: Andreas.Kendziorra@DLR.de

M. Weber
e-mail: Melanie.Weber@DLR.de

© Springer International Publishing AG, part of Springer Nature 2019
M. Behrisch and M. Weber (eds.), *Simulating Urban Traffic Scenarios*,
Lecture Notes in Mobility, https://doi.org/10.1007/978-3-319-33616-9_6

oped within the project VABENE [2]. Until recently, this system considered only motorised individual traffic. Due to the fact that not only this means of transport is affected by an extraordinary occurrence, the EmerT system was expanded by more means of transport to allow multimodal considerations. As SUMO [3] is one part of this system, it clearly had to be extended as well. Public transport (bus, tram and train) and pedestrians were already a part of SUMO prior to the expansion, however in a basic fashion. Therefore, the implementation was enhanced to achieve more realistic simulations. Furthermore, new functionalities enabling the simulation of logistic transport in SUMO were implemented. Moreover, since rail traffic is an essential mode of transport for public transport and logistics, rail signals were implemented to make rail-bound traffic more realistic.

The implemented extensions will be presented in this paper in the following order. In Sect. 2, the extensions regarding public transport will be presented. The implementation of the logistic concept will be explained in Sect. 3, and the one of rail signals in Sect. 4. In Sect. 5, some examples using the presented extensions will be shown. Finally, the conclusions and an outlook will be given in Sect. 6.

2 Public Transport

2.1 Existing Concept

So far, simulating public transport within SUMO was possible, however, with limited capability (see [4, 5]). The most necessary concepts already implemented for this kind of simulations were *persons*, *vehicles* and *stops*.

Persons are objects that can *walk*, use a vehicle (*ride*) and *stop*. Each person must have a plan which is a sequence of these stages. For walking, a person is following a given sequence of edges. Thereby, the walking behaviour depends on the chosen *pedestrian model*. The most advanced pedestrian model implemented in SUMO so far is the model "striping", which is a 2D model that enables walking side by side on sidewalks and includes a collision avoidance algorithm for persons that walk towards each other [6]. For a ride, a person has a list of *lines* (vehicles can be assigned to lines which can be used particularly for buses, trams and trains) and *ids* of vehicles to use. At the beginning of a riding stage, a person is positioned at an edge. Whenever a vehicle that corresponds to a line or an id of the given list stops at this edge, the person will board the vehicle. When the transporting vehicle stops at the destination edge of the ride, the person will leave the vehicle and will proceed with the next step of its plan. Stopping can be used to simulate activities such as working, shopping or doing sports. For that, a duration or a fixed end time, as well as a lane, have to be defined. For example,

```
<routes>

  <person id = "1" depart="0">
    <walk from="edge1" to="edge2" departPos="10.0"
    arrivalPos="20.0"/>
    <ride from="edge2" to="edge3" lines="a"/>
    <walk from="edge3" to="edge4" arrivalPos="30"/>
    <stop lane="edge4_0" duration="20"
    startPos="40"/>
    <ride from="edge4" to="edge5" lines="b"/>
  </person>

</routes>
```

Vehicles have the capability to transport persons. For that, there existed no restrictions on how many persons a vehicle can transport. Moreover, there was no duration for a person to board a vehicle and any number of persons can board a vehicle in one time step. Also, there were no requirements regarding the distance between the vehicle and the boarding person. The only requirement was that both have to be positioned at the same edge.

There exists a very particular procedure of departure for vehicles, called *depart triggered*. A depart triggered vehicle has no depart time but will depart when a person boards the vehicle. This can be used to simulate parking vehicles.

As stated above, vehicles can be assigned to lines. This is in particular useful for buses, trams and trains. For example, buses that have the same route can be assigned to one line. Therefore, one can tell a person to use a bus of a certain line without choosing a particular bus. Moreover, it is possible to define *flows* of vehicles. That means, vehicles that have the same attributes, except their id and their depart time, are inserted periodically with a fixed frequency into the simulation. In particular, bus and tram lines can be defined via flows.

As already mentioned, persons can have stops as stages of their plan. However, stops can be used for vehicles and routes in general as well. In these cases, one can define further attributes. Amongst others, the start and end positions at the lane can be defined. That means, a vehicle will stop within this interval. A more advanced concept of stops is a *bus stop*. The difference is a more elaborated approaching behaviour of the vehicles (mostly buses) towards the bus stop.

2.2 Person Capacity and Person Number Extension

To resolve the unrealistic fact that a vehicle could transport an unlimited number of persons, the parameters, *person capacity* and *person number*, have been introduced. Person capacity is a vehicle-type parameter that specifies how many persons (excluding an autonomous driver) a vehicle can transport. For every vehicle class, there exists a default value for the person capacity (see [7] for a list of the default values), but it can also be set in the definition of a vehicle type (e.g.`<typevClass="passenger"id="pkw1"personCapacity=4"/>`). The person number states how many persons (excluding an autonomous driver) a vehicle is actually transporting. If the person number of a vehicle is equal to its person capacity, no further person can board this vehicle.

2.3 Boarding Behaviour Extension

Some modifications regarding the boarding behaviour of persons were implemented. As already mentioned, a person can board a vehicle only if the person number of the vehicle is smaller than its person capacity. Moreover, some restrictions about the distance of the vehicle and the boarding person were incorporated. If a vehicle is stopping at a stop and a person wants to board the vehicle, the person's position has to lie within the stop. That means, the person's position on the lane has to be larger than the start position and smaller than the end position of the stop. For depart triggered vehicles, a person positioned outside of the stop of the vehicle can still board the vehicle if the person's position has a distance of at most 10 m to the vehicle.

Another enhancement is the implementation of the vehicle-type parameter *boarding duration*. This parameter states how long it takes one person to board the vehicle. Only one person can board the vehicle at a time. Therefore, if n persons want to board one vehicle for an $n \in \mathbb{N}$, and $t \in \mathbb{R}$ is the boarding duration of the vehicle, the time required to let all persons board the vehicle equals $n \cdot t$. For example, if 12 persons waiting at a bus stop want to board a bus whose boarding duration equals 0.5 s, it takes 6 s until the last person boarded the vehicle. If the duration of the boarding of all persons exceeds the stop duration of the vehicle, the stop duration will be extended by the necessary amount of time.

Due to the changes regarding the boarding behaviour and the implementation of the parameters, person capacity and person number, scenarios using public transport can be simulated more realistically. For example, bottlenecks due to low capacities of vehicles (e.g. buses and trams) or extended travel times due to long boarding durations can be considered or may be identified in simulations.

3 Logistic

A concept for freight and goods was implemented to enable the simulation of goods traffic. This concept consists basically of the new objects called *containers* which can be used to represent goods of all kinds. For example, one can represent an ISO container, a tank container, an arbitrary amount of bulk material, an arbitrary amount of animals, etc., as a container.

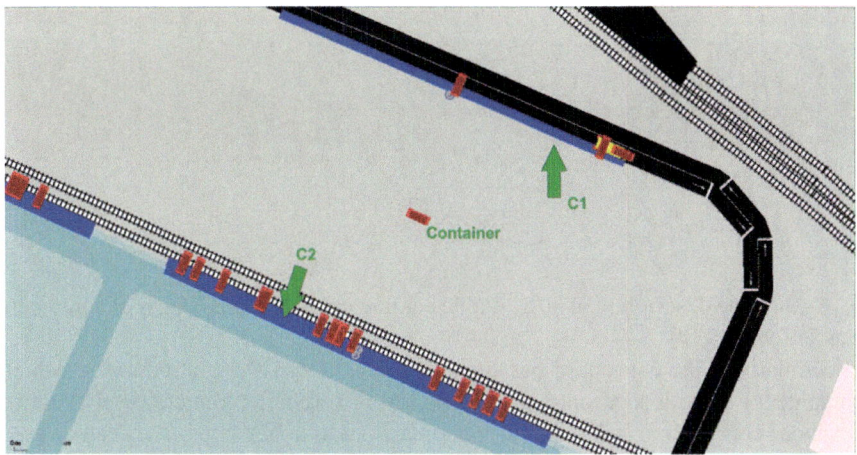

Fig. 1 A container gets transhipped from container stop C1 to container stop C2; thereby, the container moves straight from its depart position on C1 to its arrival position on C2

The concept of containers is very similar to the one of persons. Containers are objects that can be transported by a vehicle (*transport*), *stop* and that can be transhipped between two places (*tranship*), e.g. to simulate the transhipping of a container from a rail station to an adjacent harbour (which can be represented as a stop at a waterway). The mode *transport* is equivalent to the mode *ride* for persons, and *stop* is equivalent to *stop* for persons. *Tranship* defines a direct transhipping of containers between two points. Thereby, containers do not move along edges. They move in a straight line with constant velocity, no matter if the line is crossing buildings or anything else (see Fig. 1). Therefore, the time required for a tranship stage depends on the Euclidean distance between the two points and the chosen velocity (default velocity is 5 km/h).

As with the concept of persons, containers must have a plan which is a sequence of these stages. For example,

```
<routes>

  <vType id="DEFAULT_VEHTYPE" sigma="0"
  containerCapacity="1"/>
  <container id="container0" depart="0">
    <tranship from="edge1" to="edge2" departPos="80"
    arrivalPos="55"/>
    <transport from="edge2" to="edge3"
    lines="train0"/>
    <tranship from="edge3" to="edge4"
    arrivalPos="30"/>
```

```
  <stop lane="edge4_0" duration="20"
  startPos="40"/>
  <transport from="edge4" to="edge5"
  lines="truck0"/>
 </container>

</routes>
```

A complete description of all available parameters of the three stages of containers can be found at [8].

Similarly to the concept of bus stops, *container stops* were introduced at which containers can be loaded onto or unloaded from a vehicle. Vehicles use the same advanced approaching behaviour at container stops as at bus stops. They can be used to simulate transhipment stations, harbours and other places for transhipping and storing containers/goods.

To enable a realistic loading behaviour, analogously to the new boarding behaviour of persons, the parameters, *container capacity*, *container number* and *loading duration*, which correspond to the parameters *person capacity*, *person number* and *boarding duration* for persons, were introduced.

4 Rail Signals

As rail-bound traffic is an essential part of public transport and logistics, it was intended to make this kind of traffic more realistic by implementing rail signals. Rail signals are node types for controlling rail traffic. Essentially, they are traffic lights, however their state does depend only on the occupation of certain sets of edges, so-called *blocks*, instead of a traffic light programme.

More precisely, if N and M are nodes with rail signals and $L = (l_1, \ldots, l_n)$ is a connected path from N to M, where $l_1, \ldots, l_n$ are lanes, and there are no further nodes with a rail signal on this path, then L is called a *block*. The logic of the rail signals shall guarantee the following points:

- Each block of a rail network is occupied by at most one train.
- If several trains approach the same junction/railroad switch from different lanes and want to enter the same lane, at most one of these trains gets a green signal even before one of these trains has entered the targeted lane (see Fig. 2).

Due to the latter requirement, it will be avoided that trains approach a rail signal with high velocity and receive suddenly a red signal without having enough space to stop before the rail signal.

We will first explain the usage of rail signals in Sect. 4.1 before we describe their logic in detail in Sect. 4.2.

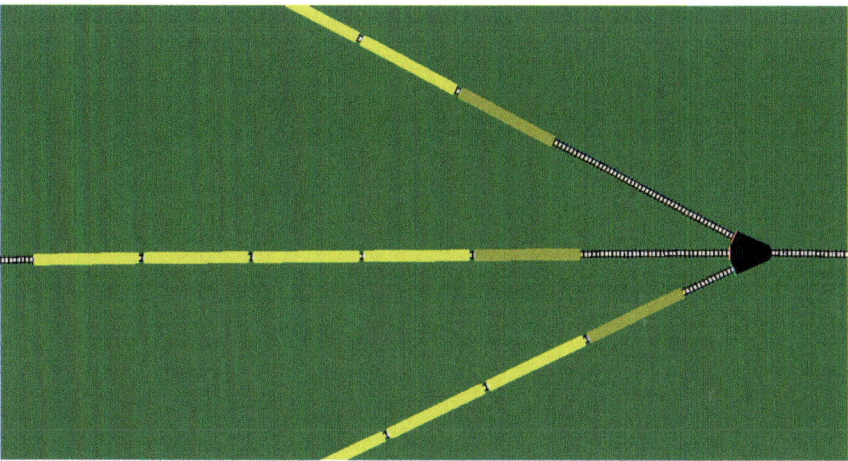

Fig. 2 Three trains approach the same railroad switch; only one has a green signal and the other ones will stop before the signal

4.1 Usage

As stated above, rail signals are node types. Therefore, if one wants to use rail signals in a simulation, one simply has to define the corresponding nodes to have the type "rail_signal", as in the following example:

```
<nodes>
  <node id="1" x="-500.0" y="0.0" type="rail_signal"/>
  <node id="2" x="+500.0" y="0.0" type="rail_signal"/>
</nodes>
```

It is worth mentioning that the usage of rail signals is not required when having rail networks, i.e. networks of edges that allow only trains, in SUMO. However, the usage of other node types, e.g. "priority", "traffic_light" or "unregulated", will result in unrealistic behaviour, which gives rise to the use of rail signals.

If one uses nodes of the type "rail_signal", some rules have to be followed. More precisely, if a node N of a rail network has a rail signal, it is required that every node that has a least three lanes (incoming or outgoing) and for which there exists a path from N has to have a rail signal as well. Otherwise, a warning will be issued and accidents and other unexpected and undesired behaviour may occur (see Fig. 3).

Rail sections without railroad switches (i.e. nodes with at least three lanes) in between can be split into several blocks by defining nodes in between to have rail signals. In reality, block lengths depend on several factors, amongst others, the line

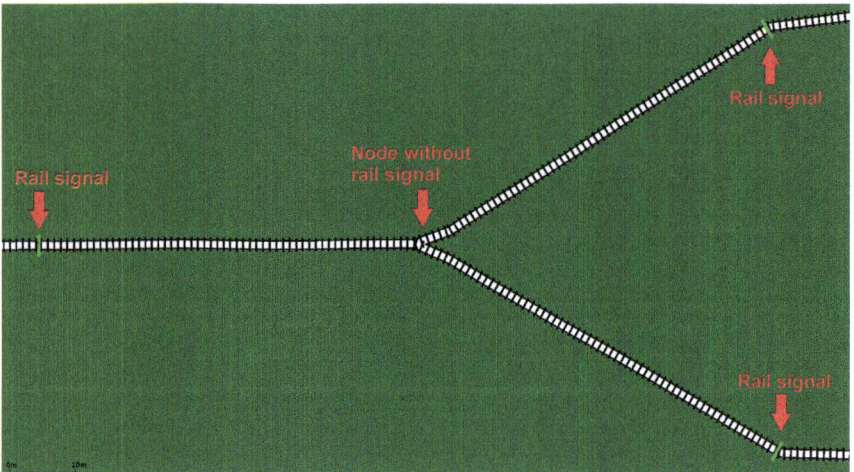

Fig. 3 A rail network containing nodes with rail signals and a node with three lanes without a rail signal; using this network in SUMO would issue a warning and could result in undesired behaviour

speed and the braking distance of the trains. Large block lengths can result in long waiting times for trains waiting to enter a block. Too short block lengths can result in trains following each other very closely. Distances of a few hundred metres can be reasonable.

4.2 Behaviour

A rail signal of a node N indicates green or red for each connection of the node individually. A connection c consists of one incoming lane i_c and one outgoing lane o_c. The signal for a connection c depends on the occupation of the following blocks:

- The outgoing block, which is the block leading from N over o_c to the next node with a rail signal. Note that, due to the requirements stated in Sect. 4.1, this block is well defined and unique.
- All blocks leading to o_c. A block leading to o_c is a block that is leading to N and whose last lane is connected to o_c (i.e. there exists a connection d at the node N, such that i_d is the last lane of the block and $o_d = o_c$). Clearly, the block to which i_c belongs to is one of the blocks leading to o_c. We denote the set of the blocks leading to o_c by O_c.

If there is only one block leading to o_c, namely the one i_c belongs to, the signal of c is green if and only if the outgoing block of c, namely the one o_c belongs to, is not occupied by a train.

If there is more than one block leading to o_c, i.e. $|O_c| > 1$, we consider all connections to the lane o_c. These connections correspond exactly to the blocks in O_c, and we denote the set of these connections by C. Moreover, we denote the set of connections in C whose corresponding afferent blocks are occupied by a train by D. We consider the following distinction of cases:

- If the outgoing block of c is occupied by a train, then the signals for all connections in C are red.
- If the outgoing block of c is not occupied by a train and all blocks leading to o_c are not occupied, i.e. $|D| = 0$, then the signals for all connections in C are green.
- If the outgoing block of c is not occupied by a train, and at least one block leading to o_c is occupied by a train, i.e. $|D| > 0$, then exactly one connection in D has a green signal and all other connections in C have a red signal.

5 Examples

5.1 Public Transport

A scenario involving persons and public transport was used to test the enhancements regarding public transport. More precisely, it was examined if unfavourable relations between the amount of people using public transport and the capacity of the public transport system affect the state of the traffic. As expected, one could find situations in which bus stops get crowded due to low frequencies of bus lines such that the buses cannot transport the amount of people intending to ride a bus of this line (see Fig. 4). This results clearly in longer travel times for persons.

Other expected consequences are congestions due to blocking buses at bus stations. This occurs when many people board a bus such that the bus blocks the corresponding lane for a long time. In Fig. 5, a situation can be seen at which 80 people board a bus with a boarding duration of 0.5 s. Consequently, the bus blocks the right lane for 40 s, which results in congestion on the right lane, as all cars intending to turn right are blocked.

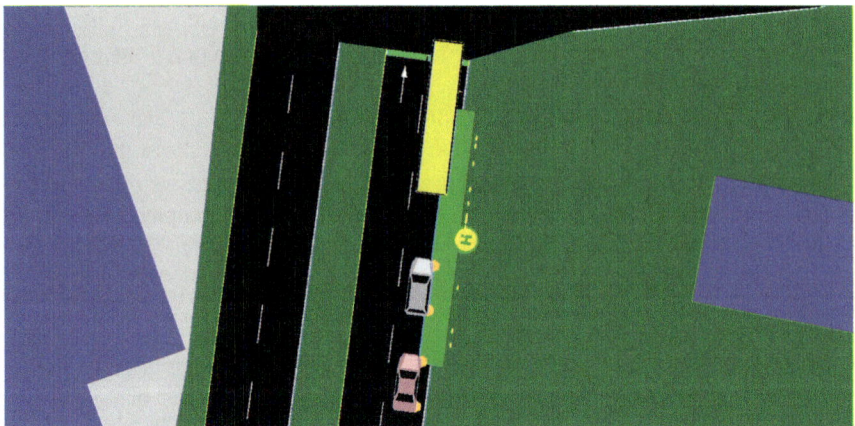

Fig. 4 A fully occupied bus leaving people behind at a bus station due to a too low capacity of the bus line with respect to the amount of people; travel time of people left behind extends accordingly

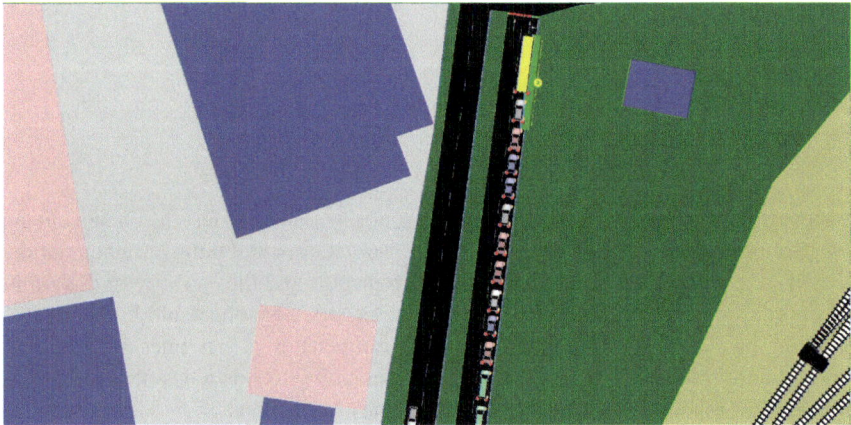

Fig. 5 Eighty people boarding a bus cause the bus to block all right turning vehicles for 40 s, resulting in a congestion on the right lane

5.2 *Logistics*

A scenario was developed to test and demonstrate the functionalities of the logistics extensions. The scenario considers the goods traffic around the harbour of Brunswick, Germany. Goods are transhipped, stored and transported by trains, ships and trucks within the scenario. An overview of the harbour area of the scenario can be seen in Fig. 6.

The harbour consists of seven landing stages, and for each landing stage there exists a corresponding goods station for trains, as well as a container stop for trucks next to the road. In this scenario, almost all the goods are transported by a vehicle

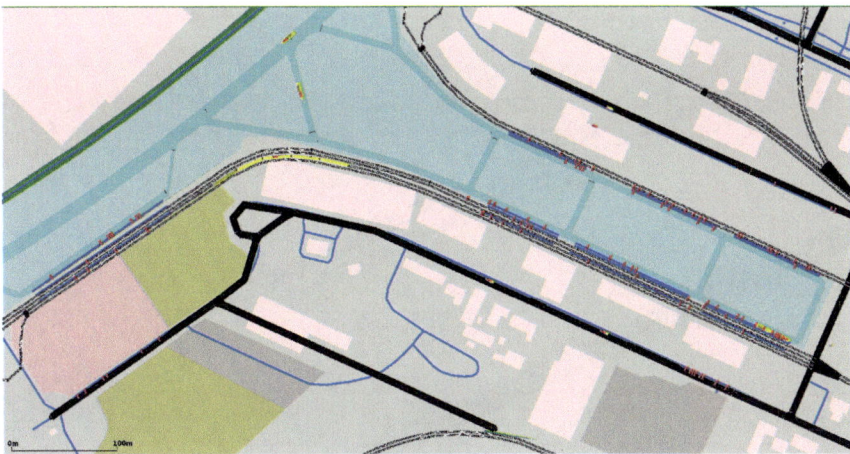

Fig. 6 Harbour of Brunswick in the SUMO simulation; containers are displayed in red, vehicles in yellow and container stops in blue

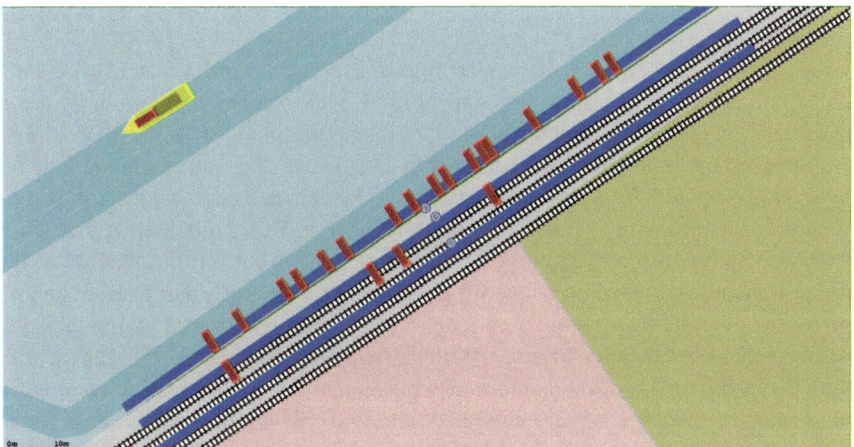

Fig. 7 Stored containers at a landing stage (container stop) of the harbour

(truck, ship or train) to the harbour area, where the container gets transhipped to another container stop, gets stored for a while and finally transported by another vehicle (see Figs. 7 and 8).

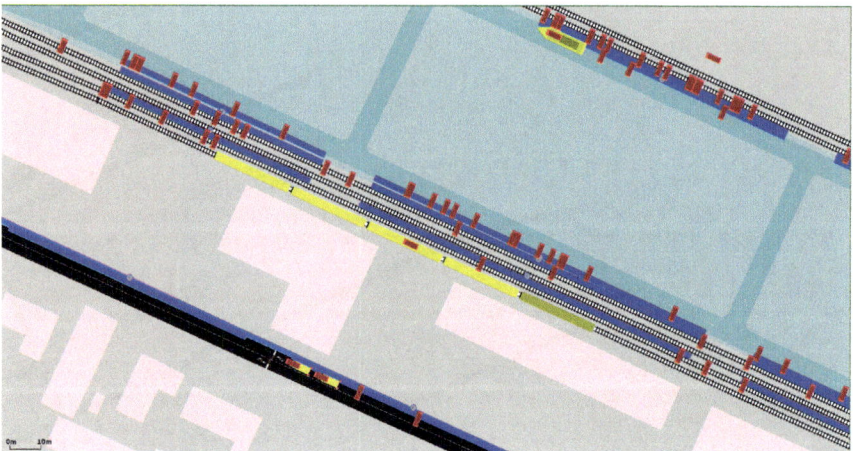

Fig. 8 Containers get loaded/unloaded onto/from ships, trains and trucks

6 Conclusion and Discussion

It was aimed to expand the existing public transport concept and to implement a first version of a logistic transport concept in SUMO to create the possibility to set up simulation scenarios with a multimodal transportation system. Furthermore, it was intended to implement rail signals for making rail traffic—an essential mode for public transport and logistics—more realistic. The enhancements allow simulations of public transport with a realistic boarding behaviour. Buses block the traffic when many people board them, and travel times of persons extend if the capacity of the public transport system is too low. In addition, major events and disaster involving persons can be better simulated. For example, in the case of a major accident, replacement buses are often inserted. With the help of the simulation, one can identify bottlenecks due to a too low number of replacement buses. The implementation of public transport reached a quite elaborated state, and only less essential extensions are thinkable. For example, an intermodal router for passengers could be implemented.

The implemented concept for freight and goods enables the simulation of goods traffic. The concept is very similar to the extended public transport concept. Due to the implementation, intermodal logistic chains can be realised now. For the future, several extensions are possible. These include a goods router or the enhancement of the goods concept, such that not only containers but also pallets or individual items can be transported. One could also enable that certain goods, such as food, bulk or liquids/gases, can be transported only by special vehicles. Furthermore, one could enable restrictions for the route of vehicles transporting hazardous material or heavy load.

The rail traffic was made more realistic due to the implementation of rail signals. A rail network can now be partitioned into blocks, whereby the blocks are separated by the rail signals. Then, a block will be occupied by at most one train, which is one of the main principles of rail traffic.

It can be summarised that the extended SUMO is able to simulate logistic and public transport which was proven in example scenarios. Further expansions are possible, however, mostly with respect to logistic as the implemented concept for public transport is quite advanced.

References

1. Behrisch M, Erdmann J, Krajzewicz D (2010) Adding intermodality to the microscopic simulation package SUMO. In: MESM 2010, Alexandria
2. Behrisch M, Bieker L, Erdmann J, Krajzewicz D (2011) SUMO - simulation of urban mobility: an overview. In: Proceedings of SIMUL 2011, the third international conference on advances in system simulation, Barcelona
3. Bundesministerium des Inneren, Nationale Strategie zum Schutz Kritischer Infrastrukturen (KRITIS-Strategie) (2015). https://www.bmi.bund.de/SharedDocs/downloads/DE/publikationen/themen/bevoelkerungsschutz/kritis.html[Zugriff am 30 5 2018]
4. Erdmann J (2013) Multimodalität und Nachfragegenerierung mit SUMO. In: Workshop Intermodalität, Berlin
5. Erdmann J, Krajzewicz D (2015) Modelling pedestrian dynamics in SUMO, preprint
6. G. A. Center, SUMO wiki – container. http://sumo.dlr.de/wiki/Specification/Containers. [Zugriff am 25 03 2015]
7. German Aerospace Center, VABENE++ traffic management for large scale events and disasters. http://www.dlr.de/vabene/en/desktopdefault.aspx. [Zugriff am 4 2 2015]
8. German Aerospace Center, SUMO wiki. http://sumo.dlr.de/wiki/Vehicle_Type_Parameter_Defaults. [Zugriff am 4 2 2015]

Part II
Scenarios

Accurate Vehicle Simulation in Logistic and Manufacturing Planning

Stefan Roth, Franz-Joseph König, Christian Dirschl and Marek Heinrich

Abstract This paper outlines the need of accurate vehicle and traffic simulation in the field of logistic and manufacturing planning. Requirements, existing solutions, and their enhancements will be described, and their abilities will be discussed. Finally, the results of bringing together the planning software suite MALAGA with SUMO, in order to provide a logistic planning suite enabled with accurate vehicle movement, will be described.

Keywords MALAGA · SUMO · Digital manufacturing · Logistic planning
In-house logistic · Congestion

1 Introduction to Simulation in Logistic Planning

Planning of production systems and their logistic supply systems originates as one of the traditional disciplines and strongholds in the automotive industry. Developments in the past have led to the inevitable capabilities of digital manufacturing. Digital manufacturing focuses on merging available information from the description of manufacturing processes and factory layout information. Yet being widely accepted in the field of production planning, methodologies of digital manufacturing also emerge into material flow and logistic planning.

S. Roth (✉)
Volkswagen AG, ITP Digitale Fabrik Logistik- und Behälterplanung, Brieffach 1832,
38436 Wolfsburg, Germany
e-mail: Stefan.Roth@Volkswagen.de

F.-J. König · C. Dirschl
ZIP Ingenieurbüro Industrieplanung und Organisation, Wolfratshauser Straße 288,
81479 Munich, Germany
e-mail: Franz-Josef.Koenig@ZIP.de

M. Heinrich
German Aerospace Center, Rutherfordstraße 2, 12489 Berlin, Germany
e-mail: Marek.Heinrich@DLR.de

© Springer International Publishing AG, part of Springer Nature 2019
M. Behrisch and M. Weber (eds.), *Simulating Urban Traffic Scenarios*,
Lecture Notes in Mobility, https://doi.org/10.1007/978-3-319-33616-9_7

Continuous effort in developing standardized, high-performing, and user-friendly software for layout-based logistic and material flow planning has been carried out though the last ten years, leading to remarkable results. For instance from the cooperation between BMW, Daimler, VW and ZIP Industrieplanung has emerged the standard planning suite MALAGA, which dedicatedly incorporates the needs of the automotive industry and provides versatile bidirectional interfaces, e.g., between the CAD System Microstation and the Simulation tool kit Plant Simulation. Layout-based logistic and material flow planning utilizes technologies such as CAD and simulation; advanced visualization capabilities of the engaged software is a key factor, flanking and driving the planning process.

Simulation studies have emerged as an indispensable investigation method in the field of industrial planning and layout development. Simulation studies in manufacturing planning vary over a broad range in their detail granularity.

Logistic layout planning in the automotive industry provides concepts and solutions for the organization of material transportation within manufacturing sites, e.g., planning how many vehicles are needed to reliably supply an assembly station in time with the needed manufacturing components.

As logistic planning evolves early in the process of projecting new facilities, when only few reliable data is available, 'static' design models are frequently used for evaluating the required transportation capacity and effort. Nevertheless, simulation studies are used, typically in the later phases of planning for testing and evaluating the developed layout.

Ongoing developments urge to bring the simulation into the planning process, becoming a constant tool in the workflow of a modern industrial planer. Enriching the static approach by ad hoc simulation studies, running in the background, validating dynamically the logistic model unveils planning discrepancies betimes. System-inherent effects such as traffic densities, congestions, and their recovering must be part of such a system.

When simulating the logistical network of a manufacturing site, often simplified models for vehicle movements are applied, due to the high modeling and computation effort and even more due to the lack of available and suitable vehicle and network modules in the leading industrial simulator software. These simplifications usually do not sufficiently consider the dynamic interactions between vehicles and with their environment.

When substituting the shortfall of embedded vehicle modules in industrial software, the desired module should provide a set network elements such as multi-lane road segments, controlled and uncontrolled intersection elements in arbitrary shapes, and with complex turning relations. The behavior of the vehicle types must be realistic/natural and be compliant with standard traffic rules. Further, it has to accurately resemble vehicle movements and driving maneuvers in pure car following or overtaking modes even when overtaking encounters intersections.

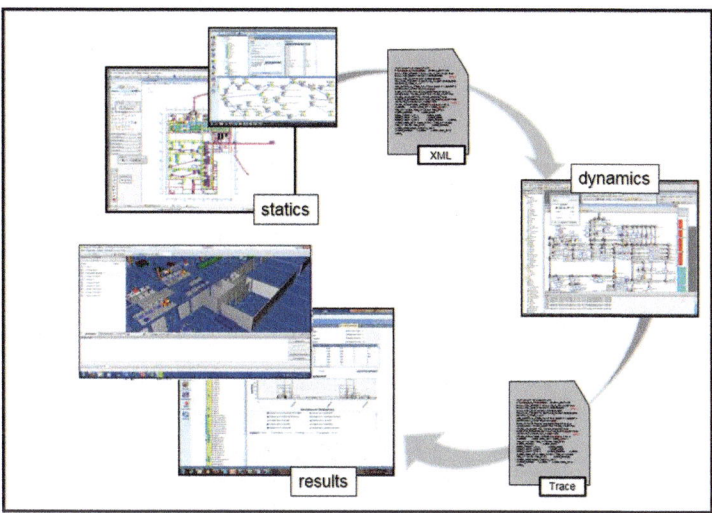

Schematic Work Flow in MALAGA

With the simulation of traffic dynamics, logistic planning gains the opportunity to identify and investigate (potentially) congested traffic and blockades due to an overload or partly a blockade in the transportation network.

Investigations on accurate vehicle movement in logistic simulations carried out by ZIP, VW, and DLR have fostered over the last three years to the development of an experimental software solution which leading to a deployable prototype in 2014.

2 Simulation Tools and Software Landscape

This paper is based on the developments that have been undertaken in a consortium of industrial partners and a research organization using the logistic planning and evaluation software suite MALAGA.

MALAGA is a planning tool kit suite assisting the logistic planner to design the in-house logistic system and material supply of a manufacturing site. MALAGA has continuously been developed since 1990. MALAGA is a distributed software suite, enabling planners to cooperatively develop their projects. MALAGA's core software components are backboned by an object database (Oracle) server operating both in online or in mobile mode.

The user interface is integrated into a CAD software for layout design. MALAGA's strength is to visualize the material flow of a production system, its interdependencies, and planning-related analyses directly in the CAD layout.

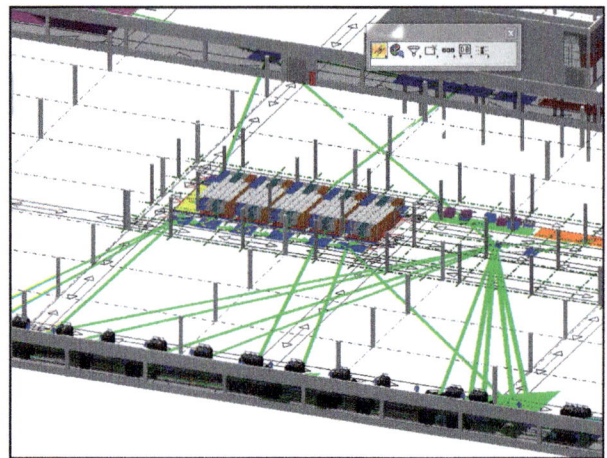

Relations of Material Flow

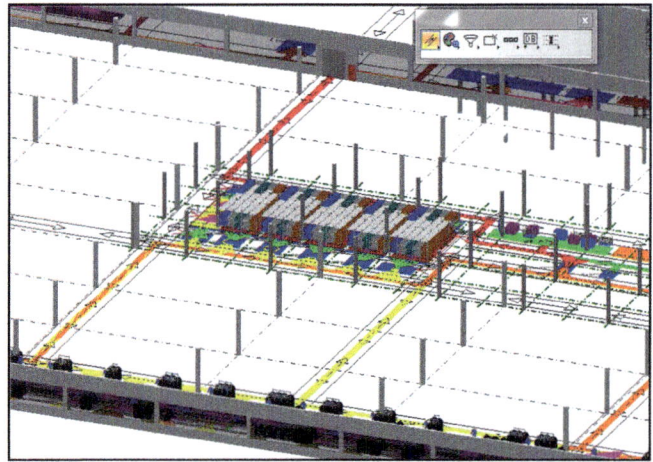

Type-specific Segment Occupation

When setting up an in-house logistic system with MALAGA, the required data must be imported first. Since MALAGA is integrated into a CAD software, existing facility layout documents and floor plans can be easily be utilized.

The layouts shall be augmented with path elements and floor space information, e.g., about warehouses and productions facilities. The corresponding elements are placed directly into the CAD layout.

To assess the material flow relations, information about the quantitative structures can be imported from a broad range of supported BoM management system. Additionally, MALAGA needs process information, which can be either automatically imported or manually amended.

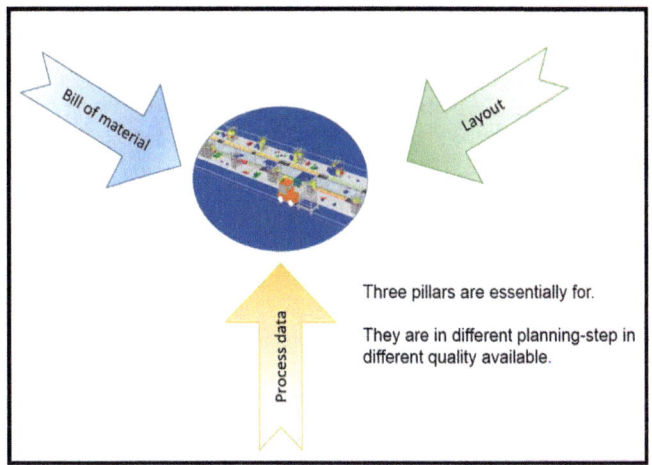

Information types used by MALAGA

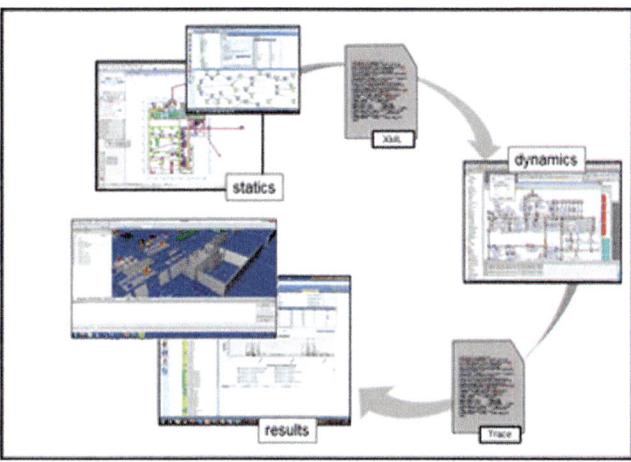

Planning Workflow in MALAGA

Together with the layout information and the quantity structure, the material flow relations can be assessed and are displayed in the layout by arrows of different strength indicating the relation's volume.

Importing floor layouts or BoM into MALAGA is a rather automatic process, whereas defining processes, especially logistical processes, is the design task of the planner shaping and analyzing the new in-house logistic system.

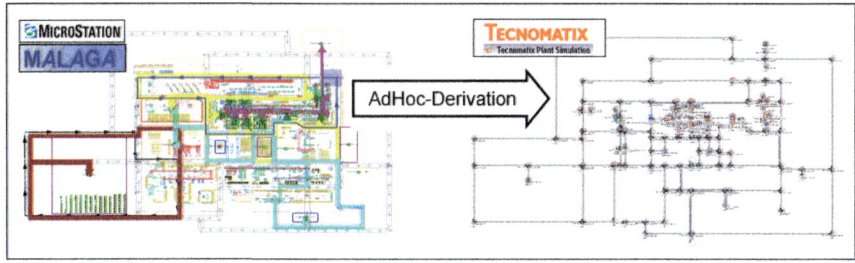

Converting a facility layout into a simulation model

For logistic supply planning, MALAGA breaks the variety of processes down into two main classes, indicating whether the process belongs to manufacturing entity (rectangular icon), or whether the process belongs to the logistic system (circular icon), e.g., certain parts need to be transported to the next machining center.

As it is inevitable important for logistic planning to know where parts are needed spatially, every manufacturing process is linked to a corresponding element (area) in the CAD layout to define its geographic position.

Yet the transport processes must be assigned. Information sets about transportation processes consist of the transport action type, transport vehicle type, its route, part handling properties, and others.

MALAGA assigns shortest paths between manufacturing stations and warehouses when routing vehicles, respecting restrictions such as transport type restriction, one-way limitations, or geometric properties of the path.

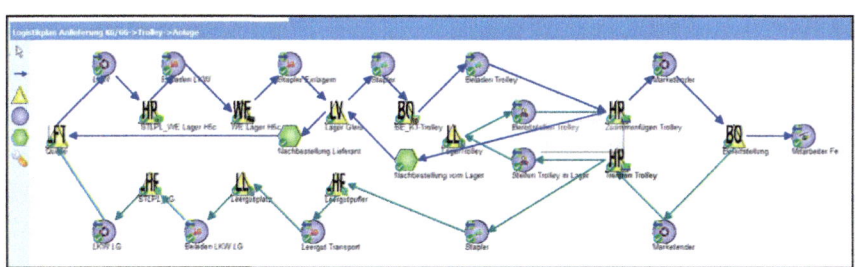

Process Chart in MALAGA

Based on the available data, MALAGA computes automatically the material flow and its characteristics. The results such as the visualization of the material flow, resource utilization and occupation rates for rolling stock and for the stores, and area consumption can easily be displayed within the layout in a 3D fashion.

When calculating the material flow, MALAGA performs both a static estimation of the material flow, based on statistical analyzes and heuristics and a set dynamic simulation studies, coping with the dynamic nature of logistic and manufacturing processes.

Therefore, since the early days, MALAGA is co-powered by a simulation engine, Plant Simulation, formerly known as Simple ++ by AESOP. The simulation framework of the MALAGA Planning Suite is customized to run on the same data.

MALAGA automatically builds up the simulation model from the available data and performs the simulation runs autonomously in the background. Relevant events and data generated during the simulation are tracked and stored for post-processing. Generally, MALAGA's simulation module is used in a computational manner and no interactive user input or specific simulation knowledge required. After terminating the simulations, the results are evaluated by MALAGA and information, e.g., on the stock inventory fluctuation and order management can be visualized.

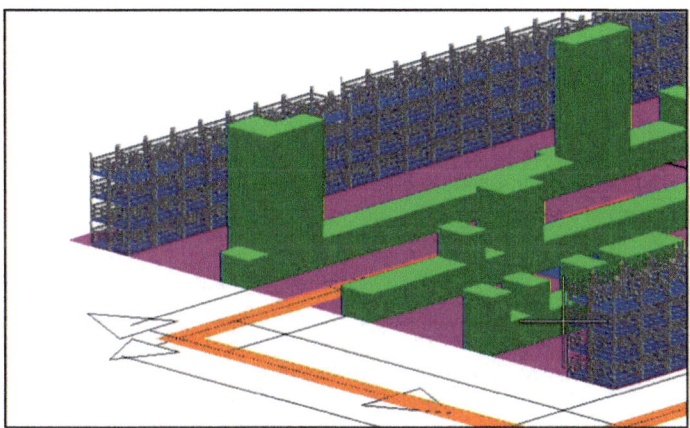

3D visualization of a Warehouse's occupation

The outcome of the planning step, the intermediate layout of the material flow system is directly visualized in the CAD layout, showing the transport network's load, the material flow volume, and the impact of measurement onto the logistic grid, highlighting stress zones and bottleneck areas for improvement in the next planning iteration.

3 Challenges in Embedding Aspects of a Road Traffic Simulation

Investigations such as [1] and their use cases show that there has always been a drive of integrating 'real-world traffic' in material flow simulation models. Often Plant Simulations' internal interpreter language is used when developing an in-model traffic simulation.

Yet no commonly used, widely spread, rich functionally equipped and even more, fast performing in-model solution could be found. Performance issues while handling intersection states and overtaking processes may have reduce the (re-)usability of the concepts in large-scale simulation studies.

When enhancing MALAGA with accurate vehicle movement, we have preferred the traffic simulation suite SUMO over the option of developing a single use or constrained intra-model traffic simulation module, for its rich abilities in simulating accurate vehicle movements, its mature and well-tested development state, and its large user community, to be coupled to the planning tool's simulation core.

SUMO has been developed as scientific tool to simulate vehicles movements in a road network. SUMO provides fast and realistic road vehicle behavior. SUMO further provides all elements and traffic participants/user types of a real road network in accordance to most central European traffic rules.

SUMO is a testified simulation environment for simulating and analyzing traffic scenarios and traffic management interventions.

SUMO has been used in countless projects for the purpose in investigating mobility and transportation, as well as for researching traffic flow and traffic management concepts. SUMO's stronghold is the proper representation of vehicles, their interactions with other vehicles and the 'world' in a coherent simulation suite. Nevertheless, for various purposes, it is flanked with numerous tools providing solutions in traffic planning-related tasks.

Due to its flexibility and its rich interfaces, SUMO has been linked to other simulation software and science application, e.g., tapas, ns3, vtd, various driving simulators and, others. Still so far SUMO has not been used in a manufacturing context and no attempt ever, to our knowledge, has been undertaken to connect SUMO to one of the leading tools in industrial planning.

Nevertheless, it is well known that linking two simulators can lead to performance drawbacks.

4   Separation of Logical Responsibilities Between the Simulators

In the existing Logistic Planning Software Suite MALAGA, Plant Simulation is used as a plug-in to provide event-driven simulation studies. Any simulation logic concerning the manufacturing processes, such as the dependencies among parts, process order, resource occupation, and other, are handled within Plant Simulation. Plant Simulation also keeps the sovereignty over the transport control logic. The transport control logic determines which parts need transportation at which time with constrains to the current state of the manufacturing system. It reserves vehicles, assigning transportation requests to transportation capacities, directs milk run organized vehicles, and finally sends transportation units into their route or to their parking position.

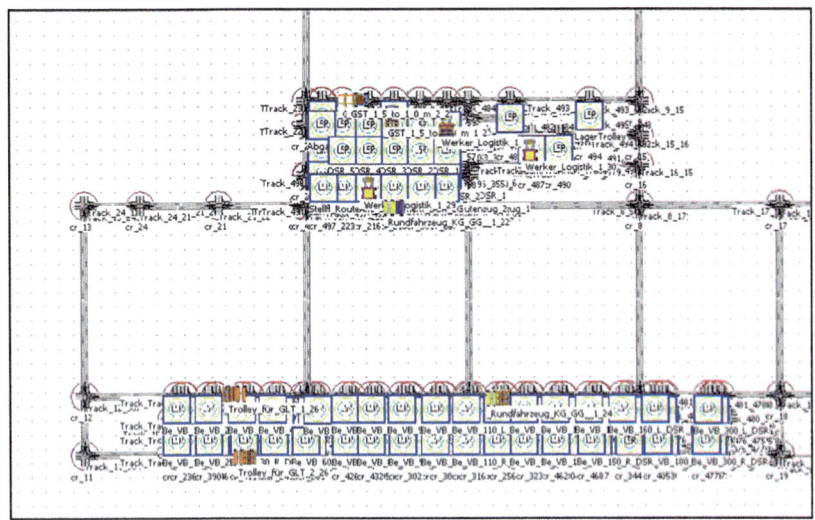

A Plant Simulation Model generated by MALAGA

When MALAGA sets up the simulation model, it also provides layout information, the configuration of the transport network and production resources, process data, and the production plan. The simulation network contains the transportation grid and the production stations, each maintaining its own registries on assembly orders, parts, and the local storage.

The simulation is performed accordingly to the production plan, parts consumption is tracked at the production stations, and once a critical amount of parts withdrawal has been reached, new part orders are released, leading to transportation requests. Transportation requests are handled by the Transport Control Module, providing and directing the vehicles in the simulation. If a transportation request had been scheduled, it must be assigned to a free transportation vehicle, fulfilling the transportation request restrictions.

MALAGA provides different, usually coexisting, transportation strategies, such as exclusive and direct fork lifter supply, transporting homogenous parts, 'milk run' supply vehicles operating on a fixed schedule, which might be carrying containers with mixed or homogenous parts, and finally taxi service vehicles, carrying order-specific containers on a predefined relation.

Depending on the transportation strategy, the transport control module gathers and merges transport requests and finally assigns new transport orders to the vehicles accordingly to the current available transportation capacities.

In unison with Plant Simulation, SUMO keeps out of any production logic, but takes over the control of any vehicle in route. SUMO simulates the vehicles' movement through network and all their interactions. For instance, SUMO receives from Plant Simulation information that a certain vehicle should move from its current position to the next manufacturing position. While moving the vehicle there, respecting traffic rules and other vehicles SUMO frequently resubmits the vehicle's motion state and its position. Traffic jams, local congestions, or speed drops related to the occupation of traffic network are handled within SUMO—natively. The effects are reported either back to transportation control, which means to Plant Simulation, or directly to the evaluation components of the planning suite.

5   Realization of the Integration of SUMO into MALAGA

The task of integrating an accurate vehicle behavior into a highly specialized, mature, and battle tested, industrial planning software suite, such as MALAGA, which is already in use at many customer sites, leaves no space for major changes in the principle software appearance, its interfaces, or its fundamental functionality. Moreover, the changes have to happen 'behind the scenes.'

The development plan for the integration of SUMO into MALAGA encountered a two-step process: from an experimental prototype up to an industry compliant MALAGA Module. The MALAGA Module, which is to be integrated into the MALAGA Planning Suite, must finally provide high reliability and performance based on mature and testified industrial technology.

The intention of the prototype phase had been to testify the technical feasibility of linking both simulators, MALAGA's Plant Simulation Module and SUMO. Within the prototype project, various technologies have been explored and methodical approaches testified in order to find the most suitable component alignment for an industry-scale integration of SUMO into the MALAGA Planning Suite.

Any crucial functionality or component of the fully featured, industry ready MALAGA Module has been evaluated in the prototype phase and software first.

The scope of this paper is to share experience made with SUMO during the experimental phase. The detail of the industry module will be presented to the community in future publications.

In order to explain how we made the common simulation of a production scenario in both Plant Simulation and SUMO possible, we would like to briefly outline the smallest version of the big picture of an integration of both simulators:

Crucial Functionality and Information Flow Capabilities:

- preparing and providing static data, which are needed in both worlds in simulator-specific formats
- initializing each simulator, with the simulation scenarios loaded
- synchronously run the simulation in both simulators
- exchange runtime-depended information between SUMO and Plant Simulation
- track, evaluate, and aggregate simulation generated information for post-analyses
- play back the relevant information to MALAGA, to be stored and visualized to the user.

Static Data—Preparing Data for Each Simulator and the Buildup of Simulation Models

Static data which has to be available in both simulators:

- Simulation settings and configuration data
- Vehicle information
- Network data and topology.

Both simulators need a scenario base for their simulation. These scenarios are specific to each simulator, following simulator-specific formats and data interpretation. Hence, it is the first indispensable premise to make the relevant data available to each simulator. Since MALAGA integrates, manages, and visualizes any data relevant for the planning of logistic processes, it is also the responsibility of MALAGA to provide the data consistently to the simulators.

Some data could be directly be diverted from MALAGA, but most data had a different meaning in each simulator and needed to be interpreted or enriched with further information.

When building common simulation scenarios, both worlds need obviously to share configuration information, just to name a few: common start time, length of SUMO's simulation steps and minimum synchronization interval or output handling directives. Truly implementing this data exchange has been uncomplicated.

Defining vehicles for both Plant Simulation and SUMO needed some enhancements of the existing data sets. Yet many vehicle information relevant for logistic planning, where available in MALAGA, such as the loading length, numbers of trailers, the maximal capacity and speed modes. Nevertheless, diver model-specific information, such as min-gap, driver's sigma a.s.o., had to be added in MALAGAS data model.

The topology of the network first appeared to be very similar, but then turned out to become a major undertaking in aligning the simulators, since the same information is interpreted differently, leading to data inconsistencies.

Both the Plant Simulation Module and SUMO are building the road network form a tree of edges and nodes, where the nodes represent the intersections between road segments. Every node is described with its coordinates and the incoming and outgoing edges.

In Plant Simulation, the road junctions are directly represented by nodes. The nodes themselves are point of intersections of edges. Especially they occupy no space because they have no physical extension. In Plant Simulation, the nodes represent the road junctions directly and the junctions remain points, more precisely they are point of intersections of edges. Especially they occupy no space since their area is zero.

For the early testing phases of the prototype, it was sufficient to replicate the Plant Simulation style road network in SUMO.

SUMO on its side applies the information about nodes and edges during the netbuilding process only. The tool netconvert out of the SUMO-Suite generates a SUMO-compliant network file, from node and edges information. Hereby SUMO estimates the shape of the road junction, taking the width of incoming and outgoing edges into account. Certainly SUMO junctions do have a physical extension which would be located on the original edge. As a result, the edge would be chopped respectively.

Even further the road network derived from MALAGA data contains many automatically generated access points for production cells, which are interpreted by netconvert as junctions. Since these nodes might be very close together, the resulting junction shapes would overlap and actually netconvert moves the edge's start and end points to build a SUMO-compliant network.

As this problem is also typical for other road data sources, netconvert has the ability to aggregate nodes to clusters. As most of the geometry of a MALAGA based road network is rather simple, the results of junction aggregation must be comprehensible to the user.

The results we found with netconvert's aggregation mechanism are outstanding for urban street topologies, but they create plenty of unjustified artifacts in a rather synthetic grid, dominated by rectangular angles, such as within industrial production sites. In the end of the netconvert process, we found a SUMO-compliant network, with edges which might be displaced unexpectedly compared to their original location known from MALAGA. Since information about vehicles and their position on an edge are essential for MALAGA's post-processing, the approach of aggregating junctions with netconvert was rejected.

The final option was to provide proper information about the junction directly to netconvert by precomputing node shapes. The downturn with this solution is that netconvert is incapable to compute junction internal relations, so again this information must also be provided to netconvert.

The last aspect in network building was to simulate multi-story factory sites, which MALAGA supports natively. Easily overlooked and typically underestimated SUMO brings by default sufficient capabilities to simulate vehicles movement in three spacial dimensions, probably as this might be of importance for motorway ramps and bridges in road networks.

Initialization of Simulators and Synchronous Simulation
Once the simulation scenarios have been set up for each simulator, they have to be prepared to run in parallel. For this, the middleware of both simulators must be launched by MALAGA. Since both simulators run in parallel but generally independently, the most central task of the development was to synchronize of both simulators. Synchronization is realized in the simulators middleware and partly in the simulators themselves. The middleware also establishes the communication channels and finally triggers both simulators to start and run the scenarios. The synchronization has been kept separately form any communication tasks.

Communication Between the Simulators
Since both simulators do depend on each other's simulation state and intermediate results, information must be submitted between the simulators. As the simulators are of very different nature, they must rely on rather generic exchange mechanisms. Communication between the simulators may slow down dramatically the performance.

The prototype has unveiled various communication technologies, which are used separately, depending on the information's reach and volume.

Big effort on boosting the simulators communication, especially on Plant Simulation side, is put on investigations in how to replace the vendor's socket interface.

Aggregate Simulation Generated Information and Their Visualization in MALAGA

Both simulators, Plant Simulation and SUMO, are designed to collect and evaluate data during the simulation run. Most of the data will be aggregated either already in the simulation or outside the simulation—some data will be processed only after the termination of the simulation run.

With SUMO, MALAGA is now capable to properly co-simulate vehicle movements adequately, and hence, traffic-related data is derived from the simulation. This information, in particular concerning the vehicles motion state and the dynamics in the road network, could not have been gained from a pure Plant Simulation model.

The effort of integrating SUMO into MALAGA has been undertaken, to evaluate the dynamic traffic impact onto an intra-logistic system. Hence, it is coherent to integrate the traffic-related results into MALAGA's visualization capabilities. The co-alignment of Plant Simulation and SUMO provides information about dynamic aspects of the vehicle flow within a production site, leading to congestions and in the worst case to disruptions or delays in the part supply of the in-house logistic system. It also provides information about the occupation of the network, its most stressed segments, and critical intersections tending to be congested. Loss time can be computed individually or globally for the entire network.

The evaluation emphasis focuses on vehicles and parts, providing capabilities to analyze delay times of single vehicles and parts, the total delay of orders, individually or grouped, allowing to analyze the trade-off between fleet size and utilization versus delay of part orders. This information is displayed in MALAGA so that the results can be analyzed by the user.

6 Expected Vehicle Behavior and Network Effects

The integration of SUMO into MALAGA targets at unveiling hidden capacities reductions of the transportation network due to dynamic traffic effects, based on accurate vehicle movement and interaction simulation. Certain traffic situations as they arise from a constrained network were replicated in MALAGA using the existing prototype.

The most fundamental constrain was to assure that vehicles should never run through each other while they are on the same track or lane. This must always be the case, regardless of the vehicles speed, the traffic density, the network topology. This can only be assured when vehicles know about each other, behave strategically and cooperatively, and even do plan their behavior into the future, considering the movement of other vehicle, even if they are on different segments.

Vehicles must properly respect traffic rules, and vehicles should never violate the right of way at intersections when other vehicles are approaching intersection, regardless weather the intersection is unregulated, controlled, or prioritized. The complexity of the intersections shape must not have any impact on the reliability and the compliance of the vehicles.

Intersections do reduce the free flow of a road network. With SUMO, this is respected in MALAGA, and we therefore find now vehicles jamming up at over-crowded intersections. On the other hand, congested intersection should never be completely blocked or dead locked by vehicles, leading to a collapse of the entire traffic system.

Because of the integration of SUMO, vehicles do now block other vehicles, if the road segment is not sufficiently broad for other vehicles to pass, e.g., when a unit is unloading in front of a manufacturing station, occupying parts of the road network. This might eventually even lead to vehicles unintendedly blocking the access to other production capacities.

Whereas SUMO changes a lot about the results that can be achieved with MALAGA, by simulating vehicles movements more detailed, the existing transport control module in Plant Simulation has undergone only minor changes. Thus, the control over the entire transport system remains the domain of Plant Simulation. The transport control directs SUMO's vehicles between manufacturing stations and warehouses, accordingly to the transportation requests. The transport control also remains the instance of controlling and managing disturbed or unoccupied vehicles.

7 Outlook and Discussion

With the development of the prototype for the integration of SUMO into MALAGA, ZIP, VW, and DLR have proven on a technical level that both simulators may sym-biotically be coupled. Yet not all obstacles have been removed, especially further investigation is needed in the field of handing unutilized vehicles and on manufac-turing specific artifacts in the network and its elements.

With the help of the prototype, many open development points, both in SUMO and in MALAGA, were elaborated, which must be addressed in future releases of the software.

Core enhancements will also affect netconvert and SUMO's core, especially when considering overtaking using the opposite lane.

The integration of the two simulators has yet proven its usefulness already in real-life consultant projects flanking the existing evaluations of milk run systems, leading to a reduction in the estimated vehicles fleet and containers.

The development of the industry focused, fully featured MALAGA Module has been started.

New features and capabilities are added successively. A release available to part-ners and associated institution will be available in late 2015, whereas the development will not stop and be continued in 2016.

References

1. Rizzoli AE, Montemanni R, Lucibello E, Gambardella LM (2007) Ant colony optimization for real-world vehicle routing problems. Swarm Intell. Springer 1(2)
2. Juan AA, Rabe M (2013) Combining simulation with heuristics to solve stochastic routing and scheduling problems. In: Dangelmaier W, Laroque C, Klaas A (eds) Simulation in Produktion und Logistik - Entscheidungsunterstützung von der Planung bis zur Steuerung. HNI-Verlagsschriftenreihe, Paderborn
3. Krug W, Marz L (2010) Simulation und Optimierung in Produktion und Logistik: Praxisorientierter Leitfaden mit Fallbeispielen (VDI-Buch)
4. Mayer G, Pöge C (2013) Quo vadis Ablaufsimulation – Eine Zukunftsvision aus Sicht der Automobilindustrie, 15. ASIM Fachtagung
5. Staab T, Galka S, Klenk E, Günthner WA (2013) Effizienzsteigerung für Routenzüge - Untersuchung des Einflusses der Routenführung auf die Auslastung und Prozessstabilität, 15. ASIM Fachtagung

Flood Impacts on Road Transportation Using Microscopic Traffic Modelling Techniques

Katya Pyatkova, Albert S. Chen, Slobodan Djordjević, David Butler, Zoran Vojinović, Yared A. Abebe and Michael Hammond

Abstract This paper proposes a novel methodology for modelling the impacts of floods on traffic. Often, flooding is a complex combination of various causes (coastal, fluvial and pluvial). Further, transportation systems are very sensitive to external disturbances. The interactions between these two complex and dynamic systems have not been studied in detail so far. To address this issue, this paper proposes a methodology for a dynamic integration of a flood model (MIKE FLOOD) and a microscopic traffic simulation model (SUMO). The flood modelling results indicate which roads are inundated for a period of time. The traffic on these links will be halted or delayed according to the flood characteristics—extent, propagation and depth. As a consequence, some of the trips need to be cancelled; some need to be rerouted to unfavourable routes; and some are indirectly affected. A comparison between the baseline and a flood scenario yields the impacts of that flood on traffic, estimated in terms of lost business hours, additional fuel consumption and additional CO_2 emissions. The proposed methodology will be further developed as a workable tool to evaluate the flooding impact on transportation network at city scale automatically.

K. Pyatkova (✉) · A. S. Chen · S. Djordjević · D. Butler
Centre for Water Systems, University of Exeter, Exeter, UK
e-mail: K.Pyatkova@exeter.ac.uk

A. S. Chen
e-mail: A.S.Chen@exeter.ac.uk

S. Djordjević
e-mail: S.Djordjevic@exeter.ac.uk

D. Butler
e-mail: D.Butler@exeter.ac.uk

Z. Vojinović · Y. A. Abebe · M. Hammond
UNESCO-IHE, Institute for Water Education, Delft, The Netherlands
e-mail: Z.Vojinovic@unesco-ihe.org

Y. A. Abebe
e-mail: Y.Abebe@unesco-ihe.org

M. Hammond
e-mail: M.Hammond@unesco-ihe.org

© Springer International Publishing AG, part of Springer Nature 2019
M. Behrisch and M. Weber (eds.), *Simulating Urban Traffic Scenarios*,
Lecture Notes in Mobility, https://doi.org/10.1007/978-3-319-33616-9_8

Keywords Microscopic traffic modelling · Road networks · Traffic disruption
Model integration · Flood modelling · Flood impacts

1 Introduction

Floods can impact human activities in many ways, and this is why it is common
to categorize these impacts. The flood consequences can be grouped as direct or
indirect, tangible or intangible or combinations of both [21]. *Direct* damages occur
if the asset of interest is physically exposed to flood waters (i.e., buildings, people
or environment). *Indirect* damages are usually outside the flooded area and usually
become apparent after a longer time [17]. A classic example of indirect losses is
the interruption of production in a firm that might occur due to a supplier affected
by flooding. Traffic disruption due to floods is another indirect flood impact, the
importance of which has not been studied in detail. The main reasons are as follows:
(1) the complexity of integrating two highly dynamic and uncertain systems; (2) the
need to assess flood impacts in monetary terms (for the purposes of cost-benefit ratio).
Flood impacts on traffic are often intangible: loss of time, frustration, environmental
degradation due to additional CO_2 emissions. However, they can also have monetary
dimensions: additional operating costs and fuel consumption have market prices, and
loss of time could be monetized as well. Approaches to monetize the intangibles and
the emerging importance of multi-criteria analysis for hazard impact assessments
create the necessary conditions for the proper evaluation of flood impacts on traffic.

To date traffic disruption due to flooding has received little attention. Comprehensive flood impact guidelines recommend carrying out traffic disruption studies
only if the expected traffic losses are significant, because otherwise the cost of traffic
disruption is negligible compared to direct or indirect tangible costs [20]. However,
the importance of impacts on traffic (relative to other flood impacts) varies—in some
cities (e.g. Beijing), flooding has exaggerated the problems in their congested transportation network; but in other cities, it is not so significant. So far, the flood impacts
on traffic have been approached using simple mathematical models [20] or macroscopic traffic models [3, 24]. None of these methods considers the dynamics of the
transportation system, rerouting whilst a street is closed, or the dynamics of the
flooding event. These methods represent a static system, which uses homogeneous
aggregated traffic flows. The reliability of such models is not high, especially when
it comes to simulating decisions in complex urban traffic networks. Microsimulation represents traffic congestion situations and bottlenecks more realistically, mainly
through its algorithms incorporating drivers' responses and intermodal transportation
[11, 14].

Hitherto microsimulation has not been used for computing flood impacts on traffic
congestion, and this is one of the main goals of this research. From the modelling
part, an evident gap in the current research is the fact that traffic models are not based
on the duration and propagation of the flood. The methods introduced in this paper
will address this dynamic behaviour of the system, through timely changes of the

status of the links (open, closed or with a certain speed limit in accordance to the changes in flood depth).

Further improvement in the hydraulic component of the methodology is also desired. The hydraulic models used by Suarez et al. [24] and Chang et al. [3] do not simulate surface flooding, caused by insufficient drainage capacity. Hydraulic modelling is a key element in urban flood management [4, 8], especially when a realistic representation of flooding on the streets is necessary.

As extreme events are the main cause of flooding, this research is also interested in the impacts of adverse weather conditions on transportation systems. There is no indication that effects of bad weather have been included earlier to model flood impacts on traffic. Heavy rainfall events lead to a reduction in speed and capacity of the network [6, 9, 12, 15, 16, 18]. Keay and Simmonds [12] found that traffic delays increase with the intensity of precipitation. Smith et al. [23] estimated that during rainfall events average speed reductions are approximately 5–6.5%. Nokkala et al. [19] used another approach to estimate travel delays due to bad weather. Speed reductions of 20, 30 and 40% were assigned to trips with different average speed, assuming that shorter trips will have lower average speed than longer trips. This assumption is valid for mathematical models, but it disregards the different road capacities on the network. For the purposes of the methodology in this paper, the speed reduction will be applied as a proportion of maximum speed limit per link in the road network. This will allow vehicles to have varying speed according to their routes.

The connection between rainfall, reduced visibility, increase in reaction time and speed reduction is apparent from drivers' safety viewpoint. On the other hand, adverse weather impacts on traffic demand are more complex, because it might lead to the cancellation of trips or changes in travel modes [1, 5, 16]. The feedback to traffic demand will be assessed in the current research by cancelling trips that have origins or destinations in the flooded area.

This paper focuses only on the methodology for integrating flood and road transportation models, and results discussion is going to be subject to future work.

2 Methodology

The proposed conceptual framework for incorporating flooding conditions into a microscopic traffic model is outlined in Fig. 1. As discussed earlier, the impact of extreme hydro-meteorological events on transportation is twofold—coming from rainfall events and flooding of the road network. First, the extreme weather conditions lead to reduced maximum speed limits [12]. As different streets in the network have different speed limits, the atmospheric conditions will define a proportionate speed reduction in each link. The decrease of speed limit will be driven by the intensity and the duration of the rainfall event, and it will reduce road capacity before the flood has even occurred. Thus, the flood impacts will start evolving in a transportation system, which already has reduced capacity due to heavy rainfall intensities.

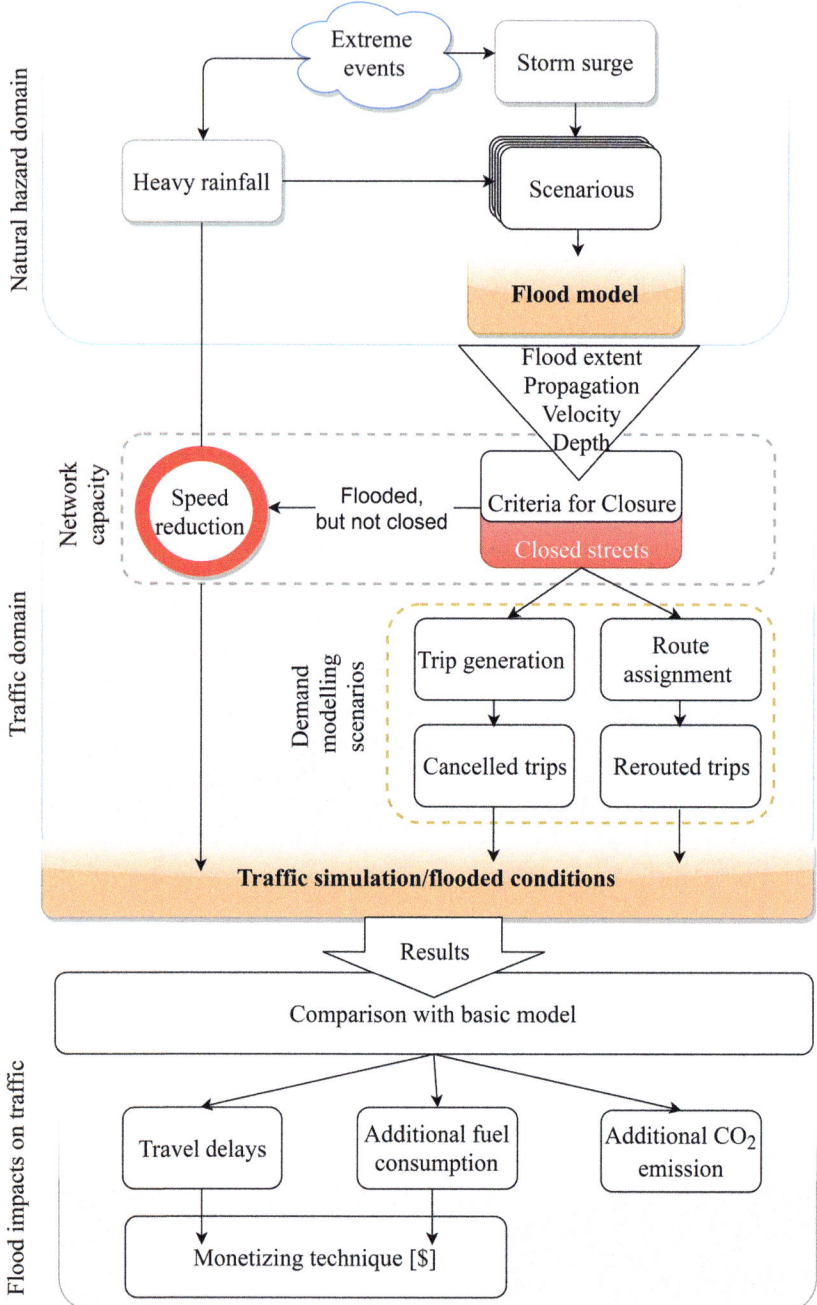

Fig. 1 Flow chart of the proposed methodology

Different combinations of intensities of rainfall and storm surges are simulated to produce the time-varying flood characteristics. The consequent flood intensities in terms of flood extent, depth and propagation determine whether a street in the road network is going to be closed for traffic. This closure will affect the overall road capacities, the trip definition and the route assignment components of the flood model. Trips having an origin or destination in the flooded area will be cancelled, and the routes that pass through a flooded area will be rerouted to unfavourable routes.

The rerouting process assumes that drivers have no initial information that a part of their route has been flooded. Their route diversion is done when they approach the link closed for traffic, and then, a new route is assigned based on the shortest path to their destination.

A microsimulation technique facilitates a better and a more detailed representation of the traffic processes, compared to macrosimulation. There are several reasons to adopt a microsimulation technique for the assessment of flood impacts:

- When a street is closed due to flooding, each vehicle will be rerouted individually, according to its destination. Hence, the rerouting algorithm ensures a detailed representation of the traffic condition during flooded conditions. This is particularly important if there are numerous flooded streets throughout the whole network;
- The microsimulation technique is more reliable for estimating losses, related to the trips that are cancelled due to flooding because it contains a detailed description of each trip and its purpose;
- The intermodal representation of different vehicle types is important for the overall consumption of fuel and greenhouse gas emissions. Different modes of transportation also indicate different cost of travel delays and will result in a more realistic representation of the flood impacts.
- Microscopic traffic models can simulate the dynamics of the flood propagation both in spatially and temporally. For instance, depending on the flood severity, it can allow closure of only one lane, whilst keeping the traffic active in other lanes;

The results of the traffic simulations will be compared for scenarios with and without flooding. The whole procedure will be performed for different flooding scenarios, different times of the day (peak and off-peak times). As stated before, the results will be presented in absolute measures of lost business hours, additional fuel consumption and additional CO_2 emissions. The travel delays and the additional fuel consumption will be also represented in monetary terms so that they can be easily compared to the other type of flood losses and damages in the studied area. Ultimately, such an approach will allow the effects of both flood risk management measures and of traffic improvement systems to be tested.

The model will be applied to a case study on a Caribbean island—St Maarten. This case study is considered appropriate for the research for two reasons: first, it has been a frequent victim of tropical storms and hurricanes; second, the closed road network system of an island makes it easier to assess indirect impacts.

The following sections elaborate the hydraulic model used to simulate the flooding conditions, and the translation of flooding results into SUMO environment and the SUMO modelling set-up.

2.1 Hydraulic Model

The case study area of St Maarten is prone to tropical storms and hurricanes. Even small-scale floods in the past posed a serious threat to traffic [27]. The insufficient drainage capacity or the lack of drainage structures results in considerable flood depth on the streets during severe storm events [27]. The hydraulic modelling has been carried out on a catchment level for the most hazardous catchments in the island of St Maarten. The flood hazard characteristics (depth and velocity) were computed using DHI software MIKE FLOOD [7]. This software ensures full dynamic coupling between MIKE 11 (1D river model) and MIKE 21 (2D model, computing the floodplain and the coastal flooding). The results from the coastal flooding model were used as boundary conditions in the MIKE FLOOD simulation. Thus, surface runoff and coastal conditions were integrated at each time step. The flooding conditions were simulated for different return periods of storm events, assuming independence of the rainfall and storm surge occurrence. The results of the hydraulic model provided maps for relevant flood depth over time, depending on the flood propagation at a particular site.

2.2 Translation of Flood Model Results into SUMO Model Input

The time-varying flood maps identify the streets that will be closed and the duration of the closure. This extraction was performed in a GIS environment by overlaying the flood map with a road network (Fig. 2). The road map is a modified version of Open Street Map (OSM), which ensures all street types and their corresponding speed limits are correct. In OSM format each street is defined by its ID, and this is not suitable for the purposes of a precise identification of flood locations. The desired level of precision requires each edge of the network to have a unique ID, which can be identified by the traffic model. In order to avoid conversion discrepancies, a reliable translation of the link indices was required. This was performed first by segregating major streets into edges in a GIS environment and by giving unique indices of the individual edges. The resultant shapefile was saved as an OSM and then translated to a network file, readable by SUMO. That way, no data will be lost in conversion, i.e. space varying speed limits, or the number of lanes per street. By ensuring same edge IDs, a linkage between the ArcGIS and SUMO environments was established. Thus, a list of flooded streets was identified by their IDs in GIS environment and was readily available for rerouting in SUMO. To provide consistency, the newly created roadmap was used for simulating traffic for the baseline scenario and for the various flooding scenarios.

The criterion for a street closure was based on flood characteristics—extent, propagation, depth and velocity. Guidelines for a street closure can be found in studies related to car stability in flood water. Stationary vehicle stability in flood waters has

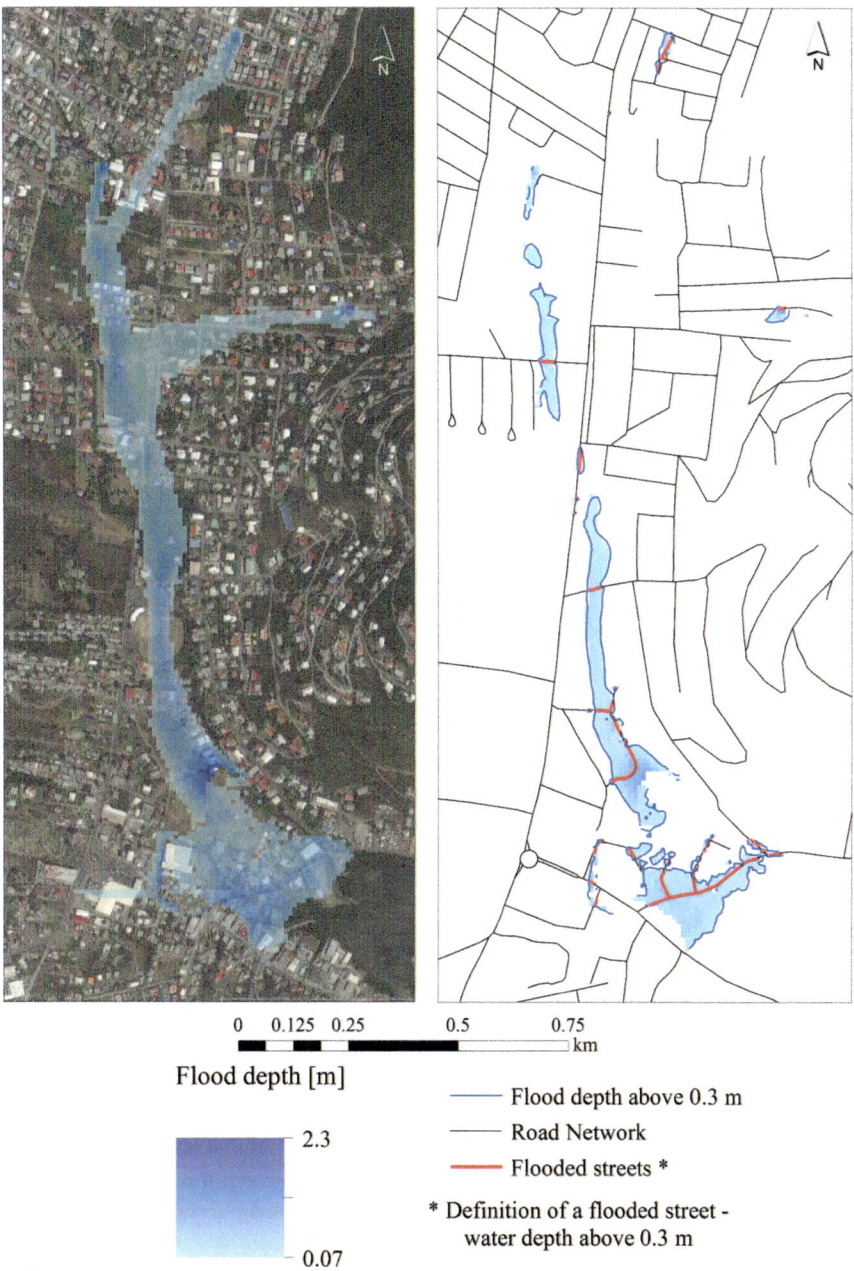

Fig. 2 Flood map with 100 years return period of a rainfall event (left), and a road map overlaid with a flood map with flood depth above 0.3 m (right), showing in red the roads closed to traffic

Fig. 3 Flow chart of the implemented traffic model

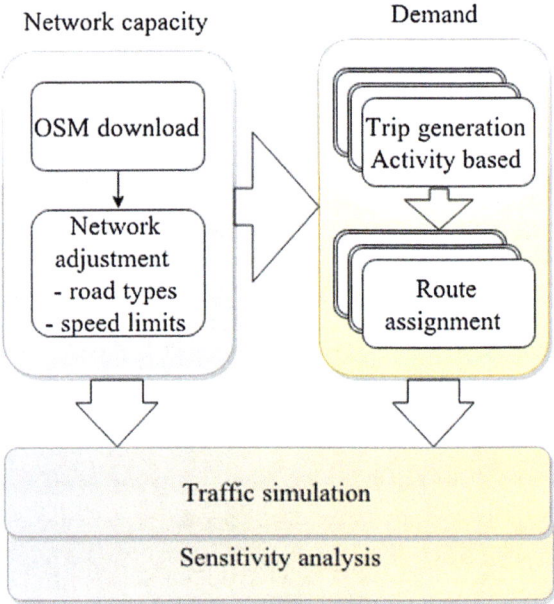

been an object of experimental [25] and analytical studies [13]. Shand et al. [22] carried out a literature review for the purpose of establishing guidelines for vehicle safety on the road. They found a considerable agreement among the literature about the floating limits of different types of vehicles—small passenger (0.3 m), large passenger (0.4 m) and 4DW (0.5 m). These guidelines were adopted in the current research as a rule for a road closure: streets with more than 0.3 m flood depth were closed for traffic.

2.3 SUMO Parameters Setting and Traffic Volume Estimation

The SUMO software [2] has been used to create a basic model, so that the proposed methodology can be tested (Fig. 3). The traffic model was limited by the reduced availability of transportation measurement data, but it is believed there are sufficient data to further test the methodology. Currently, the model uses the traffic network of the whole island of St Maarten, which is rather large for conventional microsimulation network (total area 87 km^2 and nearly 80 000 inhabitants).

This network has been extracted from OSM and later on modified for the applications in SUMO, with special attention given to the different types of roads and their maximum speed limits. The traffic demand was established on a quasi-random route generation (ActivityGen), based on statistics about settlements, population and the locations of big employers and schools. Different parameter combinations were used to run the model and obtain statistics for each scenario.

The route assignment of the current model employs shortest path algorithm to establish a route between the origin and the destination of each trip. For a better representation of route choices, an algorithm representing shortest travel time is required.

Sensitivity analysis of different scenario results can help improve the understanding of how the system functions. This strategy for computing traffic demand was hoped to help overcome the lack of data for calibration. A large number of quasi-randomly generated data approaches the traffic demand distribution from a probabilistic standpoint.

Another objective of the research is to investigate what the environmental impact of traffic congestion during floods can be. To achieve this, the SUMO model employed a simplified HBEFA classification of vehicles and their relevant CO_2 emission levels for different engines. This model also provides a description of fuel consumption for individual trips in the simulation and thus can help monetize the impacts of floods on traffic congestion.

2.4 Monetization of Traffic Delays

Previous studies in the field of flood impacts on traffic congestion [3, 24] indicated that wasted time will be the most significant impact. This imposes the need to monetize business hours lost in traffic, so that they can be compared to other tangible flood impacts such as damage to built-environment or business interruption. Value of time per individual person (driver or passenger) is defined by the purpose of the trip, mode of transportation and the type of vehicle [28]. Monetizing of travel time has been approached using wage data to assign monetary values per hour for different trip purposes—business, commute or leisure [26]. The cost of the additional travel time can also depend on the duration of the delay. Interviews showed delays longer than 30 min have higher relative costs than shorter delays [10]. This research will employ a monetizing method which will consider a UK methodology [28] to estimate costs of travel times based on assumptions on average income.

3 Modelling Challenges and Future Improvements

We have created a preliminary model to test the proposed methodology. The trial has shown several challenges to be solved. Trivial setbacks like data conversion issues are typical when integrating models with very different inputs and outputs. The rerouting mechanisms, employed to represent street closure, are originally established to simulate traffic incidents on the road. They require an individual set-up, which is not trivial, when a large number of streets have to be closed due to the flooded conditions.

For example, when 30 cm of flood depth was used as a criterion for street closure, 268 streets in the whole network were identified to be closed and the traffic through them should be rerouted. From a traffic modelling point of view, the road network was going to suffer from 268 accidents, whose temporal characteristics depended on the propagation of the flood. In SUMO terms, each of these accidents had to be represented independently and this posed a problem when multiple scenarios were going to be modelled and discussed. As mentioned previously, the main objective of the current research is the development of a tool for the integration of flood and traffic models. Such a tool will eliminate drawbacks like the file conversion or multiple file creation for different purposes of the modelling process.

Due to the lack of traffic data, modelling different vehicle classes and purposes of trips are reliant purely on assumptions. However, different vehicle classes can give valuable input, when traffic delays or cancellation of trips is monetized. This is currently represented by the activity-based traffic demand model, which generates trips according to synthetic data about population and locations of big employers, schools and shops. Similarly, modelling emergency vehicles such as ambulances or fire brigade vehicles is very important in times of disaster management. For example, the hospital in Philipsburg in St. Maarten can be reached only by one road which is prone to flooding. Evaluation of flood impacts like these is a great challenge because losses are intangible and are no longer related to the particularities of the traffic conditions.

4   Concluding Remarks

This paper presents a novel methodology for assessing flood impacts on traffic. Microsimulation traffic models have not been used yet to approach that problem, even though only a microsimulation model can capture the dynamics of both the natural and social-technological sphere. The impacts of adverse weather conditions on traffic have been studied in detail, but have never been previously incorporated with flood events. This methodology combines the joint impacts of both adverse weather conditions and accumulated floods to road transportation.

The methodology presented in this paper will be implemented in real-world applications. The actual traffic measurements will be applied to calibrate the parameters used in the SUMO to ensure the modelling results can represent the traffic conditions properly. Cost assessment model of travel delays also needs to be adjusted to regional specifications of salaries in Saint Martin.

This interdisciplinary approach relates to offline analysis of combined flood and traffic modelling. The methodology lends itself nicely for real-time modelling and decision-making for coupled flood and traffic management systems.

Acknowledgements Research on the PEARL (Preparing for Extreme And Rare events in coastaL regions) project is funded by the European Commission through Framework Programme 7, Grant Number 603663.

References

1. Al Hassan Y, Barker DJ (1999) The impact of unseasonable or extreme weather on traffic activity within Lothian region, Scotland. J Transp Geogr 7:209–213. https://doi.org/10.1016/S0966-6923(98)00047-7
2. Behrisch M, Bieker L, Erdmann J, Krajzewicz D (2011) SUMO—Simulation of Urban MObility—an overview. Presented at the SIMUL 2011, The third international conference on advances in system simulation, pp 55–60
3. Chang H, Lafrenz M, Jung I-W, Figliozzi M, Platman D, Pederson C (2010) Potential impacts of climate change on flood-induced travel disruptions: a case study of Portland, Oregon, USA. Ann Assoc Am Geogr 100:938–952. https://doi.org/10.1080/00045608.2010.497110
4. Chang T-J, Wang C-H, Chen AS (2015) A novel approach to model dynamic flow interactions between storm sewer system and overland surface for different land covers in urban areas. J Hydrol 524:662–679. https://doi.org/10.1016/j.jhydrol.2015.03.014
5. Chung E, Ohtani O, Warita H, Kuwahara M, Morita H (2005) Effect of rain on travel demand and traffic accidents. In: Presented at the 2005 IEEE intelligent transportation systems, 2005. Proceedings, pp 1080–1083. https://doi.org/10.1109/itsc.2005.1520201
6. Cools M, Moons E, Wets G (2010) Assessing the impact of weather on traffic intensity. Weather Clim Soc 2:60–68. https://doi.org/10.1175/2009WCAS1014.1
7. DHI (2007) MIKE FLOOD modelling of urban flooding, A Step-by-step training guide
8. Djordjević S, Chen A, Leandro J, Savić D, Boonya-aroonnet S, Maksimović Č, Prodanović D, Blanksby J, Saul A (2007) Integrated sub-surface/surface 1D/1D and 1D/2D modelling of urban flooding. Proc Aquaterra World Forum Delta Coast, Dev
9. Doll C, Trinks C, Sedlacek N, Pelikan V, Comes T, Schultmann F (2013) Adapting rail and road networks to weather extremes: case studies for southern Germany and Austria. Nat Hazards 72:63–85. https://doi.org/10.1007/s11069-013-0969-3
10. Douglas NJ, Franzmann LJ, Frost TW (2003) The estimation of demand parameters for primary public transport service in Brisbane attributes. In: Australasian transport research forum (ATRF), 26th, 2003, Wellington, New Zealand
11. Helbing D, Hennecke A, Shvetsov V, Treiber M (2002) Micro- and macro-simulation of freeway traffic. Math Comput Model 35:517–547. https://doi.org/10.1016/S0895-7177(02)80019-X
12. Keay K, Simmonds I (2005) The association of rainfall and other weather variables with road traffic volume in Melbourne, Australia. Accid Anal Prev 37:109–124. https://doi.org/10.1016/j.aap.2004.07.005
13. Keller RJ, Mitsch BF (1992) Stability of cars and children in flooded streets. http://search.informit.com.au/documentSummary;dn=696127213419517;res=IELENG. Accessed 1 May 2015)
14. Kerner BS, Klenov SL (2003) Microscopic theory of spatial-temporal congested traffic patterns at highway bottlenecks. Phys Rev E 68:036130. https://doi.org/10.1103/PhysRevE.68.036130
15. Koetse MJ, Rietveld P (2009) The impact of climate change and weather on transport: An overview of empirical findings. Transp Res Part Transp Environ 14:205–221. https://doi.org/10.1016/j.trd.2008.12.004
16. Maze T, Agarwai M, Burchett G (2006) Whether weather matters to traffic demand, traffic safety, and traffic operations and flow. Transp Res Rec J Transp Res Board 1948:170–176. https://doi.org/10.3141/1948-19
17. Merz B, Kreibich H, Schwarze R, Thieken A (2010) Review article "Assessment of economic flood damage". Nat Hazards Earth Syst Sci 10:1697–1724. https://doi.org/10.5194/nhess-10-1697-2010
18. Michaelides S (2014) Vulnerability of transportation to extreme weather and climate change. Nat Hazards 72:1–4. https://doi.org/10.1007/s11069-013-0975-5
19. Nokkala M, Leviäkangas P, Oiva K (2012) The costs of extreme weather for the European transpor tsystems (No. EWENT project D4)

20. Penning-Rowsell E (2010) The benefits of flood and coastal risk management: a manual of assessment techniques/Edmund Penning-Rowsell... [et al.].: Middlesex University Press, London

21. Penning-Rowsell E, Chatterton J, Rowsell EP (1980) Assessing benefits the of flood alleviation and land drainage schemes. ICE Proc 69:295–315. https://doi.org/10.1680/iicep.1980.2539

22. Shand TD, Smith GP, Cox RJ, Blacka M (2011) Development of appropriate criteria for the safety and stability of persons and vehicles in floods. http://search.informit.com.au/documentSummary;dn=317612923491163;res=IELENG. Accessed 1 June 2015

23. Smith BL, Byrne KG, Copperman RB, Hennessy SM, Goodall NJ (2004) An investigation into the impact of rainfall on freeway traffic flow. In: 83rd Annual meeting of the transportation research board, Washington DC

24. Suarez P, Anderson W, Mahal V, Lakshmanan TR (2005) Impacts of flooding and climate change on urban transportation: a systemwide performance assessment of the Boston Metro Area. Transp Res Part Transp Environ 10:231–244. https://doi.org/10.1016/j.trd.2005.04.007

25. Teo FY, Xia J, Falconer RA, Lin B (2012) Experimental studies on the interaction between vehicles and floodplain flows. Int J River Basin Manag 10:149–160. https://doi.org/10.1080/15715124.2012.674040

26. Tervonen J, Ristikartano J, Penttinen M-M (2010) Tieliikenteen ajokustannusten yksikköarvojen määrittäminen. Liikenneviraston Tutkimuksia Ja Selvityksiä 33:2010

27. UNDP (2012) Innovation and technology in risk mitigation and development planning in SIDS: Towards flood risk reduction in Sint Maarten. United Nations Development Programme, Barbados and the OECS

28. Vickerman R (2000) Evaluation methodologies for transport projects in the United Kingdom. Transp Policy 7:7–16. https://doi.org/10.1016/s0967-070x(00)00009-3

Improving Traffic Lights System Management by Translating Decisions of Traffic Officer

François Vaudrin and Laurence Capus

Abstract Coordination of traffic signal timing systems has significant impacts on traffic congestion, waiting time, risks of accidents, and unnecessary fuel consumption. Actually, systems of traffic light's programming involve complex calculations especially to tackle problematic situations in real time. Another way of doing is to manage traffic flow by traffic officers. Despite the limitation of short-term retention of human brain to few elements, human being can make decisions in case of system malfunction or during special events. The human strategy as that of the traffic officers is simple and is based on common sense. This paper explains how to implement this strategy and gives some results obtained. The simulation is performed with the open-source traffic simulation software, simulation of urban mobility (SUMO). The preliminary simulation results are promising for the continuation of this research. The observation of patterns could bring to propose an intelligent system more efficient that reuses similar cases to save time.

Keywords Microscopic traffic simulation · Open source · SUMO · Traffic lights Traffic management · Traffic research · Artificial intelligence

1 Introduction

The process of programming static traffic light's signals is complex and even more in real time. The greatest part of existing systems is electromechanical systems with clocks. The systems are programmed for fixed periods (rush hour in the morning, in the afternoon, weekend, etc.) and are not changed for several years thereafter. But

F. Vaudrin (✉) · L. Capus
Department of Computer Science and Software Engineering, Faculty of Sciences and Engineering, Pavillon Adrien-Pouliot, Université Laval Québec, 1065, Avenue de la Médecine, Québec, QC G1V 0A6, Canada
e-mail: francois.vaudrin.1@ulaval.ca

L. Capus
e-mail: Laurence.Capus@ift.ulaval.ca

© Springer International Publishing AG, part of Springer Nature 2019
M. Behrisch and M. Weber (eds.), *Simulating Urban Traffic Scenarios*,
Lecture Notes in Mobility, https://doi.org/10.1007/978-3-319-33616-9_9

even during these periods, there are fluctuations in traffic flow and it is important that the system is adapted to these variations.

The Institute of Transportation Engineers [2] (ITE[1]) recommends taking an average traffic volume during a given period and using these data to develop various programming plans with a simulation software until a satisfactory solution is obtained. The simulations are typically made by using commercial simulation software. Thereafter, it is implemented on site and revalidated by a traffic expert who makes the final adjustments required (fine-tuning). Besides that, a study conducted in more than 100 states, cities or agencies on behalf of the ITE showed that more than half of traffic systems in the USA were badly synchronized or poorly maintained [8]. If it is hard to properly maintain static systems, it is even more difficult to program real-time systems. The proposed solutions are often costly and time-consuming to implement, so little used.

The amount of parameters to be considered and the uncertainty caused by human behavior and the environment leads us to choose a workable and useful solution. Our goal is to manage traffic light's system at the right time and without complex calculation. For that, we considered the traffic officers' decisions as starting point. These decisions are made according to the common sense without a great training. Although these decisions are simple to identify, their implementation and validation require an important preliminary work. We need to simulate a network that reflects as closely as possible the reality and compare the performance of both the systems (static and dynamic). To make the simulation, we use open-source software SUMO,[2] developed by the Institute of Transport of the German Aerospace Center DLR.[3] The purpose of this paper is to explain the translation of traffic officers' decision-making and show its simulation on SUMO.

Section 2 describes the problem analysis that was conducted to our research objective. Section 3 explains the algorithm that enables the translation of traffic officers' decisions. Section 4 presents a first simulation and the results obtained. Section 5 concludes on the things that we have learned so far and gives some future works.

---

[1] Founded in 1930, ITE is a community of transportation professionals including, but not limited to transportation engineers, transportation planners, consultants, educators, and a network of nearly 17,000 members, working in more than 90 countries, http://www.ite.org/aboutite/, accessed March 25, 2015.

[2] SUMO is a free and open traffic simulation suite which is available since 2001. SUMO allows modeling of intermodal traffic systems including road vehicles, public transport, and pedestrians. http://www.dlr.de/ts/en/desktopdefault.aspx/tabid-9883/16931_read-41000/, accessed March 25, 2015

[3] DLR is the national aeronautics and space research center of the Federal Republic of Germany. Its extensive research and development work in aeronautics, space, energy, transport, and security is integrated into national and international cooperative ventures. DLR has approximately 8000 employees. http://www.dlr.de/dlr/en/desktopdefault.aspx/tabid-10002/#/DLR/Start/About, accessed March 25, 2015

2 Problem Analysis

Almost all of the existing systems are static and are programmed accordingly to time of the day (e.g., peak-hour morning or afternoon). But even during these periods, there are fluctuations in traffic flow and it is important that the system is adapted to these variations.

Miller [5] has shown that short-term retention of human brain is 7 ± 2 items. Despite these limitations, human being can manage traffic in case of system malfunction (or during special events). Our research aims to change traffic light's programming without complex calculations and at the right time (in real time). We do not try to find optimal solution every moment but to improve the situation the best as we can. Our assumption is that the summation of small improvements will have a significant effect over a reasonable period of time (few hours).

To get there, we will guide us on the work of traffic officers (Fig. 1). The human decision is simple and is based on common sense. Traffic officers try to redirect traffic to free places and to be fair with road users. We aim to use a similar strategy to develop an intelligent system that would be able to propose solutions like a traffic officer.

To our knowledge, no other research uses the concept of human traffic officers to develop an intelligent traffic management system. Sadek et al. [7] developed a knowledge-based system for identifying incidents on a highway and divert traffic accordingly. Krajzewiez et al. [4] proposed an algorithm based on jam lengths. Wannige et al. [9] developed an adaptive neuro-fuzzy traffic signal control for multi-

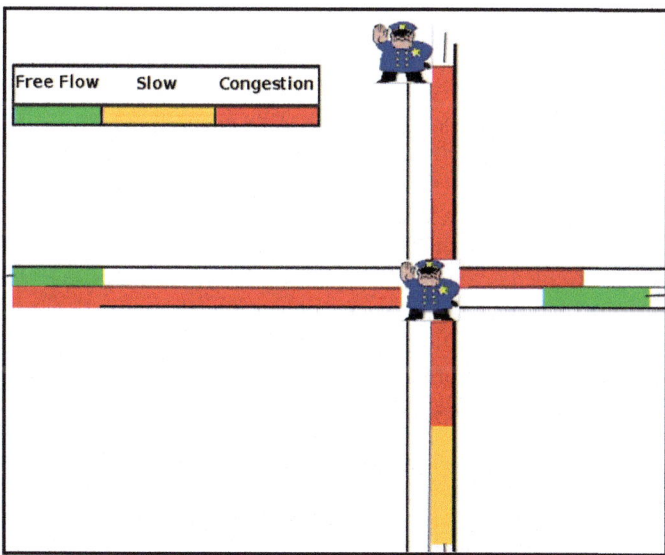

Fig. 1 Traffic officer

Fig. 2 Volume versus density (Reproduced from Pignataro—[6])

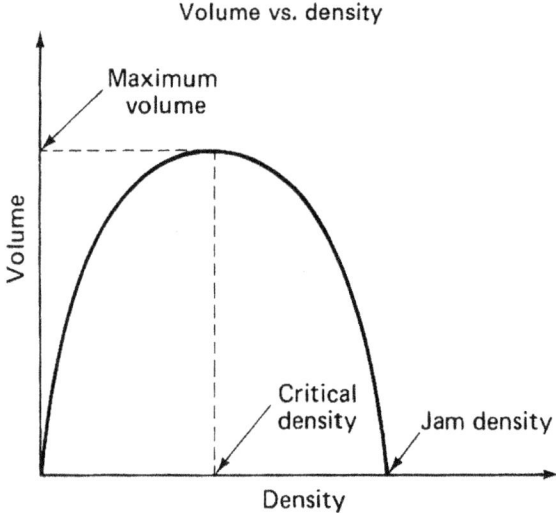

ple junctions. Hossain et al. [1] proposed a case-based system with libraries to identify recurring congestion situation on an artery with four intersections. Yang et al. [10] introduced a two-stage optimal combination fuzzy controller for traffic signals at isolated urban intersection. Kofod-Petersen et al. [3] proposed a case-based reasoning system using historical vehicle counts. Zubillaga et al. [11] proposed an approach based on entropy and complexity. These works are valuable and proposed interesting solutions but not to implement decisions done by traffic officers. Our approach is based on common sense, and we wish eventually to apply it on a complex network.

The challenge is to translate this common sense behavior in a system. We propose to use the density–volume curve (Fig. 2) and speed–volume curve (Fig. 3). These two figures show that traffic must not exceed a critical threshold because the situation will deteriorate rapidly. If we look at Fig. 2, we see that when the volume reaches the maximum critical point, the density is too high for the capacity of the road. And when it exceeds this point, the volume (or traffic flow) gradually decreases to the point of ultimate congestion (jam). This point is located at the right of the curve. This phenomenon is presented in Fig. 3 by comparing the volume of flow and speed. When traffic is paralyzed (speed = 0), we have reached the point of extreme congestion. This point is located at the origin in Fig. 3 and corresponds to the level of service E. The level of service is a standard of vehicle density of a road. The level of service A is a low vehicle density, while E is the ultimate density (the road is like a parking lot, and the traffic is jam). There are five levels of service in Fig. 3 (A–E), and our approach is to avoid levels of service D and E.

The aim is to calculate the density of each section (edge) and infer the level of service at every moment in each section of the network. Knowing the level of service of each section, the system will have to decide what is the best decision in this

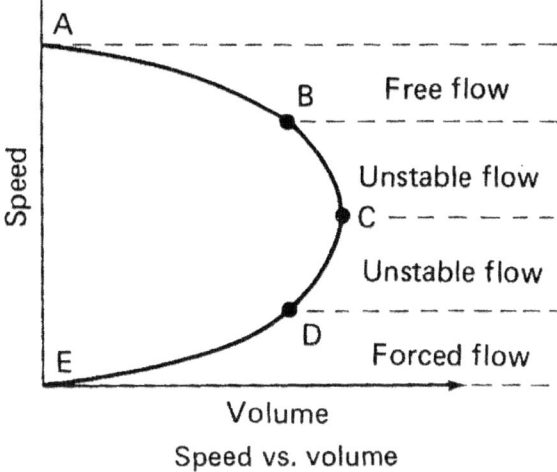

Fig. 3 Speed versus volume (Reproduced from Pignataro—[6])

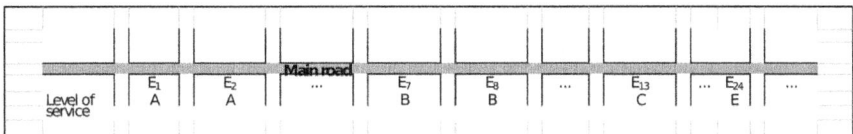

Fig. 4 Levels of service and edges

situation, as would do a traffic officer. This may result in an increase or a decrease in the duration of some traffic lights.

Let us see an example. Suppose a set of sections (edges) and a set of service levels. For each edge, the system measures the level of service. By putting these service levels one after the other, one gets a quick picture of the situation on the network. It is from this information that the decision to change the duration of traffic lights is taken.

$$\text{Edges} = \{E_1, E_2, E_3 \dots, E_n\}$$
$$\text{Level of services} = \{A, B, C, D, E\}$$

Figure 4 shows a main road with different levels of service. In this situation, the obvious decision would be to reduce the arrival of vehicles to the edges that have already been reached the level of service D or E (the last ones). A solution could be to reduce the duration of green lights for the previous edges. It will probably not solve congestion problem but avoid extending the level of congestion in previous edges, and it can reduce average waiting time for the secondary roads.

3   Translating Decision of Traffic Officer

The basic idea is to find a balance in the network as a traffic officer would do. Given the cognitive limitations of the human being, her/his decisions must be based on common sense and experience. For example, if a section has too high occupancy rate, it is possible to modify the duration of the green light or to change the programming cycle. For each problematic situation, the goal is to seek which decision will reduce the inconvenience and unnecessary waiting time. If the occupancy of a section (edge) is higher than a threshold, the system will modify accordingly the programming of the traffic light's system. Another objective is to avoid unnecessary waiting time for vehicles, like when the light is red and there are no vehicle on the cross road. This occurs in less busy times, and it is frustrating for the users.

At each intersection, there are output's edges, and the idea is to check the level of service of these edges. If the level of service of a given edge is too high, the system has to decide which decision it has to take. This decision could be to increase or to decrease the duration of the green or of the red light. It could also be to change the traffic light's system for a limited time. But before changing the duration of fires, it has also to check if the way is clear on the other edges by measuring the level of service of these edges. This is after analyzing this information that the system will decide which decision is the best, as would do a traffic officer.

The basic ingredient of the algorithm that we propose is based on the occupancy rate of section roads. If a secondary section has a low occupancy rate (e.g., 10% or minus—level A), we consider that the situation is under control and it is not necessary to modify the programming of traffic lights. If it is at D level, it means that there is a problematic situation under development and it is preferable to intervene in changing the traffic light's programming.

We use SUMO libraries and the function **getLastStepOccupancy**(edgeID), to develop this algorithm. This function transfers the occupancy rate to percentage at the last step on the given edge. SUMO also allows to increase the duration of traffic lights or to change phases with **setPhaseDuration**(tlsID, phaseDuration) or **setProgram**(tlsID, programID). The other element to consider is the ability to absorb traffic. So, before changing the duration of a traffic light system, the algorithm has also to check if the destination road sections are clear.

Finally, we have to decide the number of levels of service needed. Although there are five levels of service in Figs. 2 and 3 (A to E), we believe that it is possible to reduce them to meet the needs. However, the number of levels of service could be adjusted depending on the results. So, we decided to start with only three level of service and to use these rates:

If 10% or less of occupancy rate → A
Between 10 and 40% occupancy rate → C
More than 40% of occupancy rate → D

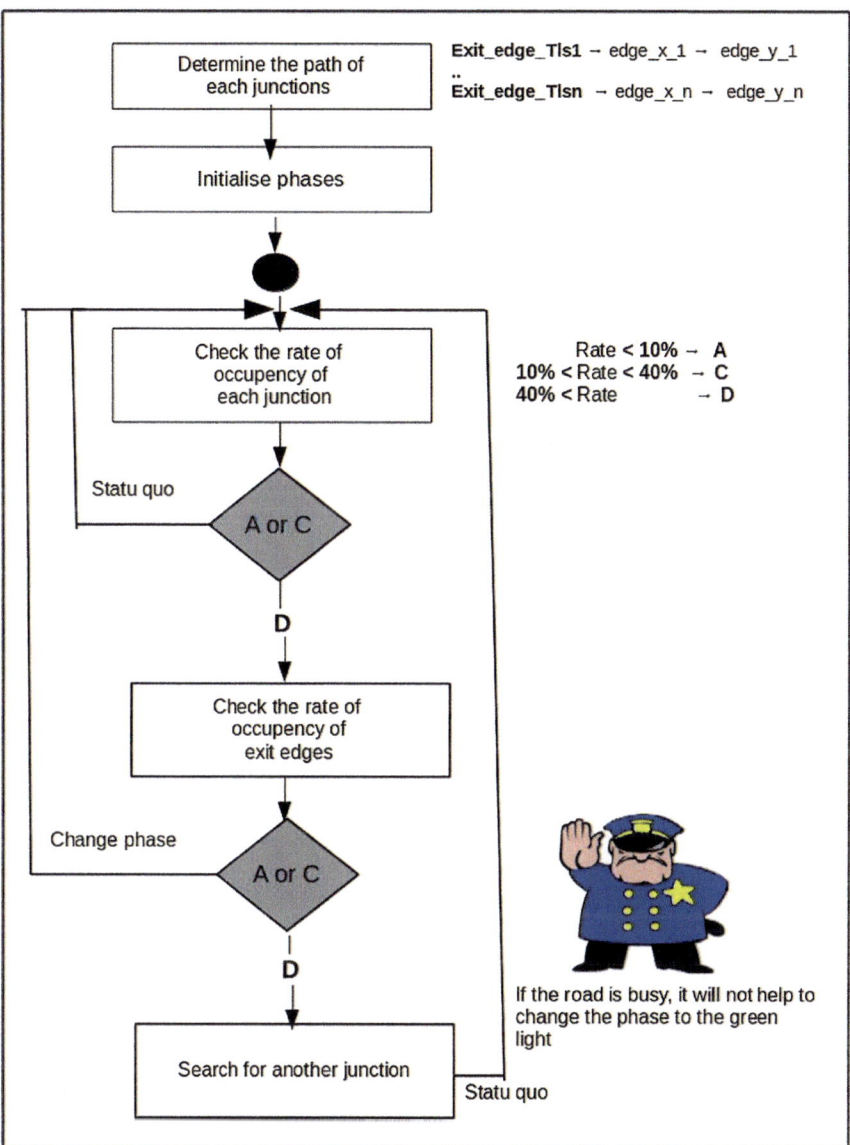

Fig. 5 Traffic officer algorithm

Figure 5 shows the main steps of the algorithm.

The traffic officer algorithm may also be summarized in this way:

1. Determine minimum service levels and thresholds that should not be exceeded.
2. Determine critical sections and sequences to be considered (e.g., the outlet section of an intersection and the next section to go).

3. Calculate the occupancy rate of each section with **getLastStepOccupancy**(edgeID) function.
4. Compare with thresholds to be respected.
5. Decide whether it is better to increase the duration of the green light or red light, or to change the phase of one or more traffic lights. To do so, use the **setPhaseDuration**(tlsID, phaseDuration) or **setPhase**(tlsID, index) or **setProgram**(tlsID, programID) of SUMO libraries.
6. Make appropriate changes and repeat the cycle.

In the next section, we present how we use SUMO to simulate a particular context in order to apply the traffic officer algorithm on this context. We also show some results obtained for this first simulation.

4 Simulation and Results

In SUMO, a street network consists of nodes (junctions) and edges (streets connecting the junctions). Thus, if one wants to create a network with two streets, subsequent to each other, he needs three nodes and two edges.[4] Moreover, SUMO allows changing signal programming plan during the simulation with TraCI module that is part of the system. Also, it is possible to make a set of traffic light's program with it and to use them at the right time accordingly to the algorithm.

The network is located in Québec City in Canada along the Sainte-Foy road (Fig. 6). This is a busy area with significant congestion problems. The studied network stretches over a distance of about 4 km. This network is made using free software, OpenStreetMap (OSM[5]) and the Java application JOSM.[6] The vehicle's paths infer from the data collected in the field. Data were measured with the short-count method. The short-count method evaluates the number of vehicles that pass each intersection for 15 min and then extrapolates the results for one hour. Each intersection was observed during the peak period of the afternoon between 16:00 and 18:30. In addition, the signaling systems in place were measured (duration of each cycle to each intersection). Geometric data were collected from the OSM map and validated in the field. By cons, in a first step, we decided to limit ourselves to a small portion of the network. We found that the task would be too complex and that it was preferable to be limited to few intersections.

During simulation, there is a continual back and forth between the screen and the XML files and it is better to easily find the information. So, to facilitate identification of nodes and edges, a Python program was carried out and IDs have been changed.

[4]Tutorials/Hello Sumo, Hello Sumo—Introduction, http://sumo.dlr.de/wiki/Tutorials/Hello_Sumo, accessed March 25, 2015.

[5]OpenStreetMap (OSM) is a collaborative project to create a free editable map of the world, https://en.wikipedia.org/wiki/OpenStreetMap, accessed March 25, 2015.

[6]JOSM is an extensible editor for OpenStreetMap written in Java 7, https://josm.openstreetmap.de/, accessed March 25, 2015.

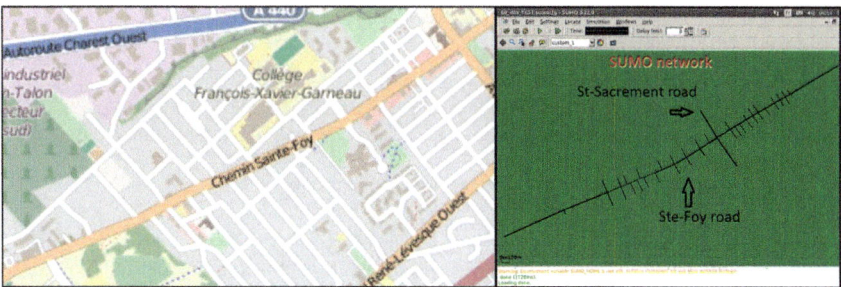

Fig. 6 OSM and SUMO

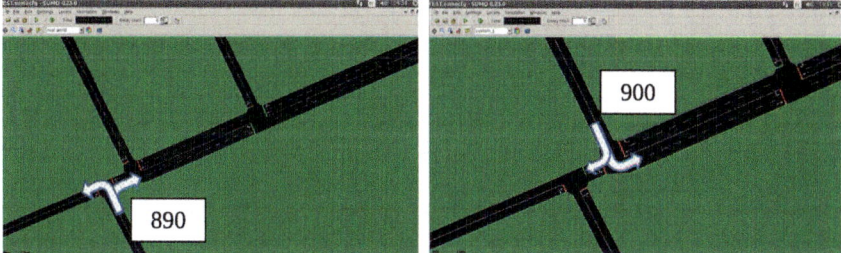

Fig. 7 Network to study

We think it is important because it would speed up the analysis. This approach also reduces the risk of errors. The nodes on the main road are now classified from West to East or left to right (100, 200, 300, 400, 500,.... 2600), and those on the secondary roads are incremented by 1000 according to the x-coordinate of each terminal node (5000, 6000, 7000,... 44000). A similar approach was used for the edges.

It seems important, before testing a new system, that the system we intend to improve is consistent at first. To test the algorithm, we started with the three junctions shown in Fig. 7. We must also consider that the two first junctions should operate in tandem because the vehicles could be trapped after exiting the intersection. There is also a bottleneck on the main road.

After a series of tests with SUMO, we found that the best sequence was the one below. Note that the first phases relate to secondary roads to facilitate system synchronization. The minimum value of the sum of waiting times of vehicles during a simulation of 2 000 s was around 80 000 s. Therefore, we target to improve this score with the traffic officer algorithm.

```
<tlLogic id="890" type="static" programID="1" offset="0">
        <phase duration="10" state="ggrrrr"/>
        <phase duration="4" state="yyrrrr"/>
        <phase duration="20" state="rrggrr"/>
        <phase duration="4" state="rryyrr"/>
        <phase duration="60" state="gGgrgg"/>
        <phase duration="4" state="ggrryy"/>
    </tlLogic>
<tlLogic id="900" type="static" programID="1" offset="0">
        <phase duration="10" state="rrrrrGG"/>
        <phase duration="4" state="rrrrryy"/>
        <phase duration="20" state="rrgggrr"/>
        <phase duration="4" state="rrgggrr"/>
        <phase duration="60" state="gGGGgrr"/>
        <phase duration="4" state="gyyyygr"/>
    </tlLogic>
        <tlLogic id="1000" type="static" programID="1" offset="0">
        <phase duration="10" state="grrrrGG"/>
        <phase duration="4" state="grrrrgy"/>
        <phase duration="60" state="gGGGGggr"/>
        <phase duration="4" state="gyyyygr"/>
    </tlLogic>
```

We have conducted a series of tests to validate the algorithm. By changing certain phases depending on the occupation rate of main sections, we managed to achieve an overall improvement, but the waiting time of some intersections is still unacceptable. We also expect to find trends by successive approximations that could help to identify solutions.

Moreover, our first observation is that the congestion on the main road occurs mainly on the first two sections. This problem is due to the bottleneck. We also discovered that when the queue gets too long, it takes much more time to clear the way because of the wave effect. It is therefore necessary to find the right sequence for the traffic lights. All these tests will fuel our decision tree and could highlight repetitive situations to help to validate the approach. When acceptable solutions will be found, we plan to store, reuse, and improve them.

5   Conclusion and Future Work

We want to demonstrate that it is possible to develop an intelligent traffic light management system based on common sense. The traffic officer algorithm described in this paper is inspired by the work of traffic officer when he/she makes a decision. This research aims to change traffic light's programming without complex calculations.

We have made tests with SUMO and the module TraCI to change the traffic light's phase dynamically during the simulation. The preliminary results are promising but not conclusive so far. The problem is that the performance of certain parts of the network is not acceptable. The waiting times are too long, and we must do additional tests. Nevertheless, we are confident that it will be possible to enhance the efficiency of the system.

However, this study shows that it is necessary to consider the context and the environment and that each case is unique. It also confirms us that it is almost impossible to apply universal solutions. Finally, in traffic management quality of solution relies heavily on the experience of the expert. So one has to be careful and even if a solution seems attractive, field validation is essential. When the approach will be demonstrated, the future work would be to build a more efficient system capable of retaining and reusing solutions quickly and to control a more complex network.

Acknowledgements We wish to thank the German Aerospace Center (DLR) of Berlin and officials of SUMO for their continued support, especially Jakob Erdmann and Michael Behrisch who make it a duty to respond quickly and clearly to user requests. We also want to thank Bruno Rémy of OpenStreetMap Quebec group for his help and valuable advice. We finally thank the Department of Computer Science and Software Engineering of the Faculty of Science and Engineering at Laval University for financial support under a merit scholarship given to this project.

References

1. Hossain S, Kattan L, Radmanesh A (2011) Responsive signal control for non-recurrent traffic congestion on an arterial. In: TRB 90th annual meeting, Washington
2. Institute of Transportation Engineers, Traffic Signal Timing Manual, Washington (2009)
3. Kofod-Petersen A, Andersen OJ, Aamodt A (2014) Case-based reasoning for improving traffic flow in urban intersections, case-based reasoning research and development. Lecture Notes in Computer Science, vol 8765, pp 215–229
4. Krajzewicz D, Brockfeld E, Mikat J, Ringel J, Rössel C, Tuchscheerer W, Wagner P, Wösler R (2005) Simulation of modern traffic lights control systems using the open source traffic simulation SUMO. In: Proceedings of the 3rd industrial simulation conference 2005, Berlin, Germany, pp 299–302
5. Miller George A (1956) The magical number seven plus or minus two: some limits on our capacity for processing information. Harvard University-Psychological Review, pp 81–97
6. Pignataro LJ (1973) Traffic engineering—Theory and practice. Prentice-Hall, New Jersey, pp 180–182
7. Sadek A, Spencer M, Spencer M (2003) Wael El–Dessouki: case-based reasoning for assessing intelligent transportation systems benefits, computer-aides civil and infrastructure engineering. Blackwell Publishing, USA
8. Tarnoff Philip J, Ordonez J (2004) Signal timing practices and procedures, state of the practice. University of Maryland, Center for Advanced Transportation Technology, Maryland
9. Wannige CT, Sonnadara DUJ (2009) Adaptive neuro-fuzzy traffic signal control for multiple junctions. In: International conference on industrial and information systems (ICIIS), pp 262–267
10. Yang W, Zhang L, He Z, Yang Y, Fang Y (2012) Urban traffic signal two-stage combination fuzzy control and Paramics simulation. IEEE explore Digital Librar, pp 771–775
11. Zubillaga D, Cruz G, Aguilar LD, Zapotécatl J, Fernández N, Aguilar J, Rosenblueth DA, Gershenson C (2014) Measuring the complexity of self-organizing traffic lights. Entropy 2384–2407

eNetEditor: A Platform-Independent Tool for Rapid Creation of Urban Traffic Scenarios and for the Optimization of Their Energy Supply Infrastructure

Tamás Kurczveil and Pablo Álvarez López

Abstract The increasing mobility and transport demand and the sinking global supply of fossil energy carriers will eventually cause a growing trend toward alternative drive concepts and the development of corresponding energy supply infrastructures. These emerging solutions and their interaction with the prevailing traffic will need to be evaluated for their optimal integration. SUMO is a preferred tool when it comes to evaluating measures in urban traffic behavior. When using SUMO, however, the creation of corresponding scenarios is accompanied by challenges in network creation and corrections, traffic demand generation and calibration. This paper presents the newly developed tool eNetEditor, which allows users the rapid prototyping of custom and calibrated traffic scenarios and their evaluation in regard to energy consumption.

Keywords Network generation · Traffic assignment and calibration · Energy consumption

1 Introduction

Current traffic is mostly driven by fossil fuels. In 2014, the number of newly registered vehicles in Germany was 3.048.507 [1]. A total of 8522 of these vehicles (0.28%) were electric vehicles [2]. The same statistic indicates that the number of electric vehicles is expected to rise in the coming years. An exponential extrapolation of these numbers yields the range of approximately 14283–27041 newly registered electric vehicles for the year 2015.

T. Kurczveil (✉) · P. Á. López
Technische Universität Braunschweig, Institut für Verkehrssicherheit und
Automatisierungstechnik, Hermann-Blenk-Straße 42,
38108 Braunschweig, Germany
e-mail: kurczveil@iva.ing.tu-bs.de

P. Á. López
e-mail: palcraft@gmail.com

© Springer International Publishing AG, part of Springer Nature 2019
M. Behrisch and M. Weber (eds.), *Simulating Urban Traffic Scenarios*,
Lecture Notes in Mobility, https://doi.org/10.1007/978-3-319-33616-9_10

139

One of the current challenges for the customers of electric vehicles is the multiplicity of charging systems and interfaces. Whereas 14622 gas stations in Germany supply conventional traffic with the adequate amount of fossil fuel [3], only 5050 publicly available electric charging points have been installed for electric vehicles in Germany until June 2014 [4]. Many manufacturers offer proprietary solutions, tying customers to a limited amount of available charging stations, such as Tesla. In other cases, countries or regions have agreed on and effected the installation of a more consistent charging infrastructure, such as CHAdeMO in Japan and France or CCS in Germany. This heterogeneous situation has resulted in charging station infrastructures of several different manufacturers and providers with varying and largely incompatible standards. In addition to this heterogeneous situation with the charging infrastructure, new developments indicate that battery-driven electric vehicles might not remain the only alternative solution for the coming years: Most of the leading automobile manufacturers have invested many efforts into the development of hydrogen fuel-cell vehicles over the past years. Toyota has introduced one of the first of these vehicles to be sold commercially starting in 2015. This further results in a comparably increasing diversity regarding drivetrain concepts.

However, even with these trends, the development of new compatible charging infrastructures is still at its beginning. This situation is a chance for the introduction of more homogeneity in the charging infrastructure and, as opposed to the location of current gas stations, its optimal operational integration into prevailing traffic situations. For this unification of the energy supply infrastructure, traffic planners will need tools in the future that take into account the energy consumption of vehicles along their routes allowing the optimal positioning of components for the corresponding energy supply infrastructure. These results can further be used for infrastructure operators to determine the amount of energy that electric vehicles are expected to gain at specific locations within the road network and how much power will be required for their sufficient supply.

This paper introduces eNetEditor: a tool with a graphical user interface for traffic planners that allows the rapid prototyping of arbitrary road traffic networks and, as opposed to most existing tools that merely focus in network editing, the creation of calibrated scenarios based on flow measurements and the evaluation of the corresponding energy consumption.

2 Concept

For the simulation of vehicles within the road network and their interaction with infrastructure objects and other vehicles, the microscopic traffic simulation tool SUMO (Simulation of Urban Mobility) is used [5]. SUMO is described by its authors as 'an open source, highly portable, microscopic and continuous road traffic simulation package designed to handle large road networks' [5]. The development of SUMO was initiated by the Institute of Transportation Systems of the German Aerospace Center (DLR), in 2001. It has evolved into a traffic simulation tool, high in features, functionality and interfaces. Even though instantiated vehicles follow a simplified behavior, traffic simulation tools like SUMO allow the realistic replication of prevailing traffic in arbitrary road networks.

Implementations were presented in [6] that newly introduced a simplified energy consumption model for vehicle objects and a corresponding declared charging station class as integral parts into SUMO. These implementations build the functional basis for the evaluation of the energy consumptions in traffic scenarios.

The application of these implementations subsequently requires a traffic scenario consisting of a road network and traffic demand. In order to instantiate calibrated traffic in arbitrary road networks, eNetEditor (initially, a MATLAB-based tool) has been developed with a graphical user interface that allows the rapid and efficient generation of road networks. Network object data was structured such that they contain the properties required by SUMO's network generator *netconvert.exe* (i.e., edges, lanes, nodes, connections), by SUMO itself (e.g., vehicle types, bus stops), and arbitrary custom parameters and their values (e.g., traffic counts or other traffic demand data) for user-defined functions. The data structure for the network creation is outlined in Sect. 3.1. After the definition of vehicles, traffic can be instantiated and calibrated from within this tool as described later in Sects. 3.2 and 3.3, respectively. Section 3.4 gives an overview on the structure of eNetEditor's modules, while Sect. 3.5 introduces the user interface.

Using traffic data and measurements from different sources, such as flows from induction loops, eNetEditor allows the generation of traffic demand. The goal is a simulation output in form of a structure that can be used to represent energy consumption of individual vehicles over time and vehicle position or along individual lanes of a road network over time. An example urban traffic scenario will be shown in chapter "Connecting Macroscopic and Microscopic Traffic Assignment" that was generated by eNetEditor, accompanied by evaluations in regard to its constituents' energy consumptions. Chapter "A SUMO Extension for Norm-Based Traffic Control Systems" subsequently presents optimization algorithms that allow to process the resulting output of arbitrary traffic simulation scenarios.

3 Implementation

This chapter explains the structure of eNetEditor's implementations. The modules are split into three parts:

1. network definition and generation,
2. vehicle definitions and generation, and
3. traffic demand generation and calibration.

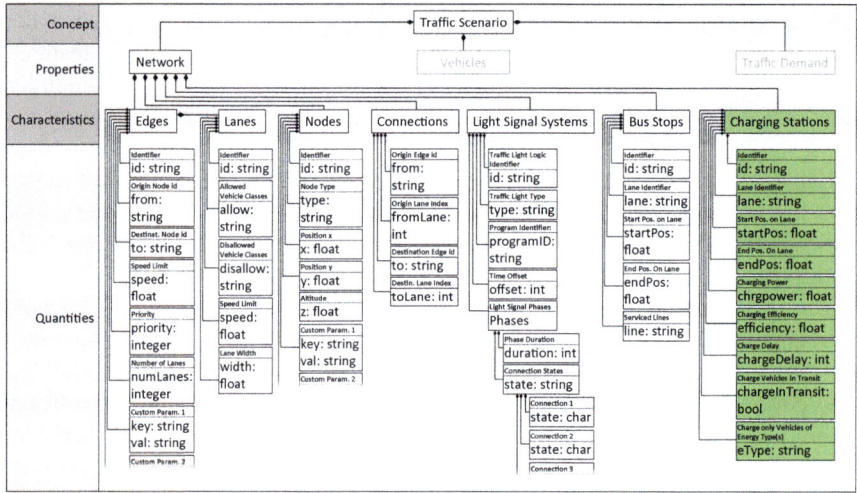

Fig. 1 Constituents of a traffic scenario's network represented in form of a UML class diagram

The following sections give an overview on the implementation of each of the modules and the used data structures that broadly represent SUMO's object structures.

3.1 Network Definition and Generation

A SUMO network consists of edges, nodes, lanes, vehicle class restrictions on specified lanes, lane connections (and prohibitions), bus stops, and light signal systems. As a new infrastructure element, charging stations have been introduced. Figure 1 shows the structure of these constituents as definable in eNetEditor, axiomatizing the terms in form of a class diagram as introduced in [7, 8]. The following sections will describe the data structure for each network constituent that can be represented in eNetEditor, the interface for their generation, and exemplified output data. Newly introduced classes and attributes are shown in green.

3.1.1   Nodes and Edges

Nodes and edges constitute the essential components of a network's graph. These can be created directly in the GUI with a mouse click or by connecting two nodes with a click-and-drag. To allow efficient analyses and operations with the resulting multi-edged directed graph and its adjacency matrix, edges and their properties are stored in the three-dimensional array E of the following structure

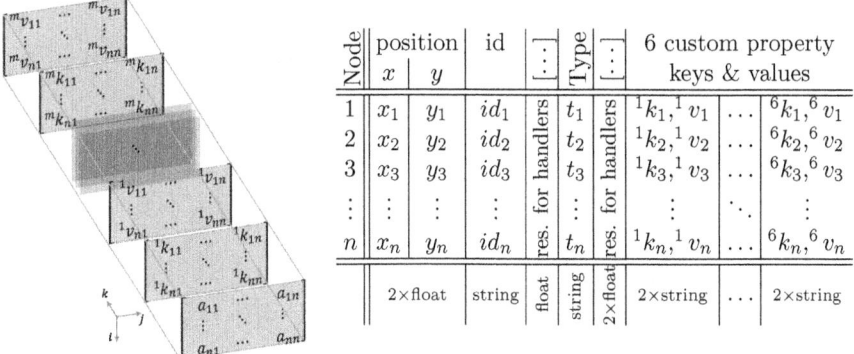

Fig. 2 Data structure of edge (left) and node (right) description arrays E and O

$$E^{n \times n \times (2m+1)} = \left(e_{i,j,k}\right), \text{ with } \dots \tag{3.1}$$

$$\text{origin node index} \quad i \quad = 1 \dots n,$$
$$\text{destination node index} \quad j \quad = 1 \dots n,$$
$$\text{property index} \quad k \quad = 1 \dots 2m + 1,$$
$$\text{number of nodes} \quad n \quad \text{and}$$
$$\text{number of edge properties} \quad m.$$

The entries $\left(e_{i,j,k=1}\right)$ represent the network graph's adjacency matrix, where each element $e_{i,j,k=1}$ equals the number of lanes from node i to node j. Edge property and corresponding value pairs are stored in successive entries of $E_{i,j,k=2,4,6,\dots}$ (property name) and $E_{i,j,k=3,5,7,\dots}$ (property value) in form of strings. Whereas SUMO and its netconvert application will only process the values of specified property tags, eNetEditor allows the declaration of arbitrary properties and property values that can be used for eNetEditor-/MATLAB-based pre-/post-processing.

Nodes and their properties are stored in the two-dimensional array O with the following structure

$$O^{n \times 18} = \left(o_{i,k}\right), \text{ with } \dots \tag{3.2}$$

$$\text{node index} \quad i \quad = 1 \dots n,$$
$$\text{property index} \quad k \quad = 1 \dots 18 \text{ and}$$
$$\text{number of nodes} \quad n.$$

Next to mandatory property values in $o_{i=1\dots n,\, j=1\dots6}$, the remaining entries in O $\left(o_{i=1\dots n,\, j=7:18}\right)$ can be used for the declaration of arbitrary node properties and their corresponding values. The structure of the edge and node variables E and O is depicted in Fig. 2.

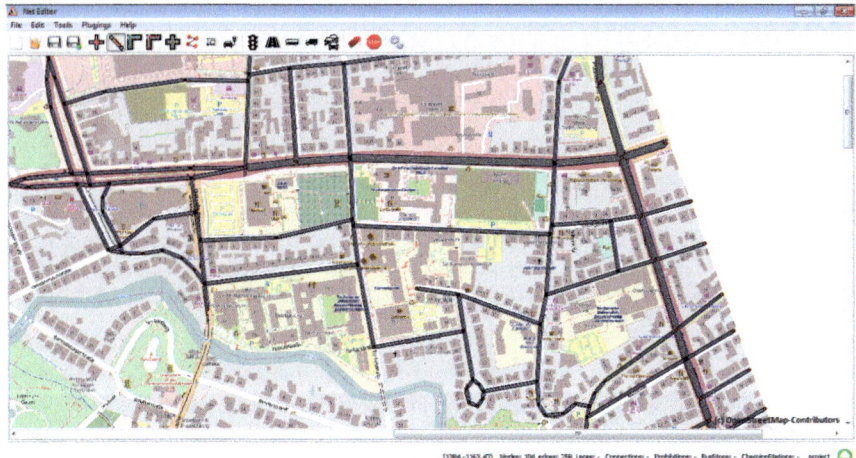

Fig. 3 eNetEditor's user interface for the creation of arbitrary road networks (c) OpenStreetMap contributors

Figure 3 shows eNetEditor's user interface for the creation of nodes and edges with arbitrary property keys and property values. The output in the specified xml format, as required by SUMO's netconvert.exe, is exemplified in Listings 3.2 and 3.1. The corresponding node and edge xml files are created by parsing the described data structures and can be initiated by a keyboard entry of the buttons *o* and *e* respectively.

Listing 3.1: Example edge definition file in the SUMO specified xml-format (typically .edg.xml)

```xml
<?xml version="1.0" encoding="utf-8"?>
<edges>
    <edge from="1" id="1to5" intensityProfile="TGw3West" numLanes="1" numVeh="
        1050" priority="4" speed="8.333" to="5"/>
    <edge from="1" id="1to11" intensityProfile="TGw3West" numLanes="1" numVeh="
        1050" priority="4" speed="8.333" to="11"/>
    <edge from="1" id="1to13" intensityProfile="TGw1West" numLanes="2" numVeh="
        14900" priority="9" speed="13.889" to="13"/>
    <edge from="1" id="1to17" intensityProfile="TGw1West" numLanes="2" numVeh="
        15050" priority="9" speed="13.889" to="17"/>
    <edge from="2" id="2to3" intensityProfile="TGw2West" numLanes="2" numVeh="
        4950" priority="7" speed="13.889" to="3"/>
    ...
</edges>
```

Listing 3.2: Example node definition file in the SUMO specified xml-format (typically .nod.xml)

```xml
<?xml version="1.0" encoding="utf-8"?>
<nodes>
    <node id="1" type="priority" x="1106.1567" y="339.0299" z="0"/>
    <node id="2" type="traffic_light" x="1074.5522" y="419.4776" z="0"/>
    <node id="3" type="priority" x="1221.0821" y="485.5597" z="0"/>
    <node id="4" type="priority" x="1313.4383" y="527.0634" z="0"/>
```

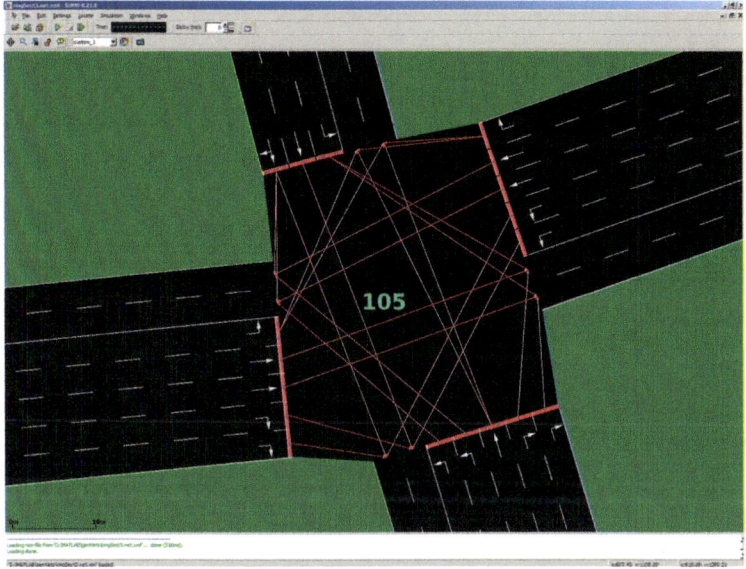

Fig. 4 A complex junction structure in SUMO with irregular connections

```
<node id="5" type="right_before_left" x="1258.4328" y="405.1119" z="0"/>
<node id="6" type="right_before_left" x="1347.5" y="443.8993" z="0"/>
    ...
</nodes>
```

3.1.2   Connections

Every edge in SUMO represents a road, on which vehicles can travel into one direction. A bidirectional street in SUMO would therefore be modeled by two edges. Each edge further consists of one or more lanes. Most nodes in a network have at least one incoming and one outgoing lane. When generating a network, netconvert will make assumptions about these connections. However, many complex junctions' structure will deviate from the assumptions made by netconvert's generalized heuristics. Figure 4 shows such an irregular junction with many specific lane-to-lane (red) connections. To allow these kinds of junction definitions and for their correct representation and functionality, connections between nodes' (junctions') incoming and outgoing lanes often need to be specified explicitly.

These connections can be specified in SUMO by explicitly declaring them in a separate connection file in xml format, typically .con.xml. Since connections are lane-specific definitions, each connection possesses six degrees of freedom: the *incoming lane*'s parameters, specified by (1) its origin and (2) destination nodes and (3) its lane index and the *outgoing lane*'s parameters, specified by (4) its origin and (5) destination nodes, and (6) its lane index. The incoming edge's destination and the outgoing edge's origin node, parameters (2) and (4), will most likely be identical.

The chosen data format for the connection specification is a three-dimensional array C, where each element $c_{i,j,k}$ represents an incoming lane and contains a list of connections to outgoing lanes. The connection variable C has the following structure

$$C^{n \times n \times l} = \left(c_{i,j,k}\right), \text{ with } \ldots \tag{3.3}$$

$$
\begin{aligned}
\text{incoming lane's origin node index} \quad & i &&= 1 \ldots n, \\
\text{incoming lane's destination node index} \quad & j &&= 1 \ldots n, \\
\text{incoming lane's index} \quad & k &&= 1 \ldots l, \\
\text{number of nodes} \quad & n && \text{and} \\
\text{lane number of the edge with the most lanes} \quad & l. &&
\end{aligned}
$$

Whereas the indices i, j, and k of each element in $\left(c_{i,j,k}\right)$ represent an origin lane, each element $c_{i,j,k}$ itself is a list of connecting destination lanes:

$$c_{i,j,k} = \begin{bmatrix} r_1 & s_1 & t_1 \\ r_2 & s_2 & t_2 \\ \vdots & \vdots & \vdots \\ r_x & s_x & t_x \end{bmatrix}, \text{ with } \ldots \tag{3.4}$$

$$
\begin{aligned}
\text{outgoing lane's origin node index} \quad & r &&= 1 \ldots n, \\
\text{outgoing lane's destination node index} \quad & s &&= 1 \ldots n, \\
\text{outgoing lane's index} \quad & t &&= 1 \ldots l, \\
\text{number of nodes} \quad & n, && \\
\text{lane number of the edge with the most lanes} \quad & l && \text{and} \\
\text{number of incoming lane's connections} \quad & x. &&
\end{aligned}
$$

Connections will be guessed by netconvert if incoming and outgoing edges exist and no lane connections are specified for a node. Connections can be specified in eNetEditor using the dialog box shown in Fig. 5, which can be called by the keyboard entry of the button c. The resulting xml file (typically .con.xml) is automatically generated after closing the dialog window. The example of a resulting xml is shown in Listing 3.3.

3.1.3 Lanes

Next to connections, lanes can be attributed with further specifications that serve two purposes: visualization and lane-based restrictions. eNetEditor allows the specifications of these introduced parameters by calling the dialog window (shown in Fig. 6) with the keyboard entry of button r. The corresponding variable R consists of a list implemented as a two-dimensional array $R = (r_i)$ with...

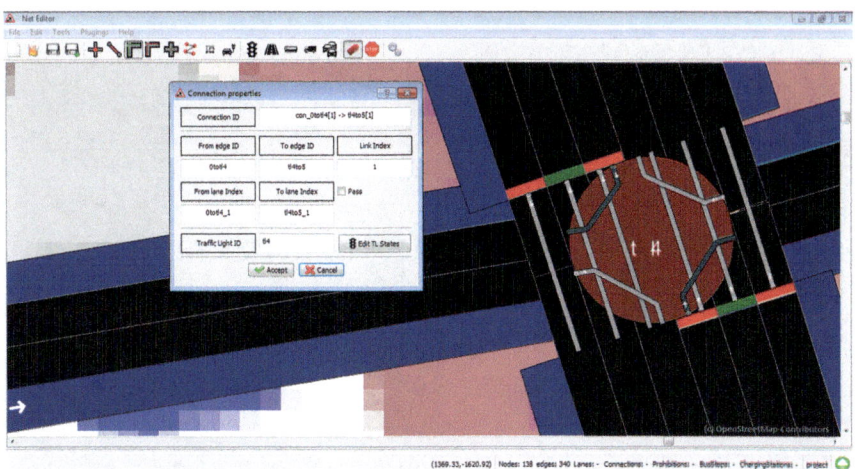

Fig. 5 Connection input dialog

Listing 3.3: Example connection definition file in the SUMO specified xml-format
(typically .con.xml)

```xml
<?xml version="1.0" encoding="utf-8"?>
<connections>
    <connection from="110to105" fromLane="0" to="105to66" toLane="0"/>
    <connection from="110to105" fromLane="1" to="105to66" toLane="1"/>
    <connection from="110to105" fromLane="2" to="105to124" toLane="0"/>
    <connection from="110to105" fromLane="3" to="105to124" toLane="1"/>
    <connection from="110to105" fromLane="4" to="105to107" toLane="0"/>
    <connection from="110to105" fromLane="4" to="105to107" toLane="1"/>
    ...
</connections>
```

$$\text{lane specification } i \quad r_i, \tag{3.5}$$
$$\text{lane specification index} \quad i \quad = 1 \dots r_{ln} \text{ and}$$
$$\text{total number of lane specifications} \quad r_{ln}.$$

Each element r_i contains all lane-based parameters for the corresponding lane:

$$b_i = \left[{}^L d_i \; a_i \; d_i \; s_i \; w_i \right], \text{ with } \dots \tag{3.6}$$
$$\text{identifier of lane } l_i \quad {}^L d_i,$$
$$\text{allowed vehicles classes on lane } l_i \quad a_i,$$
$$\text{disallowed vehicles classes on lane } l_i \quad d_i,$$
$$\text{speed limit on lane } l_i \quad s_i \quad \left(\text{in } \frac{m}{s} \right) \text{ and}$$
$$\text{width of lane } l_i \quad w_i \quad (\text{in m}).$$

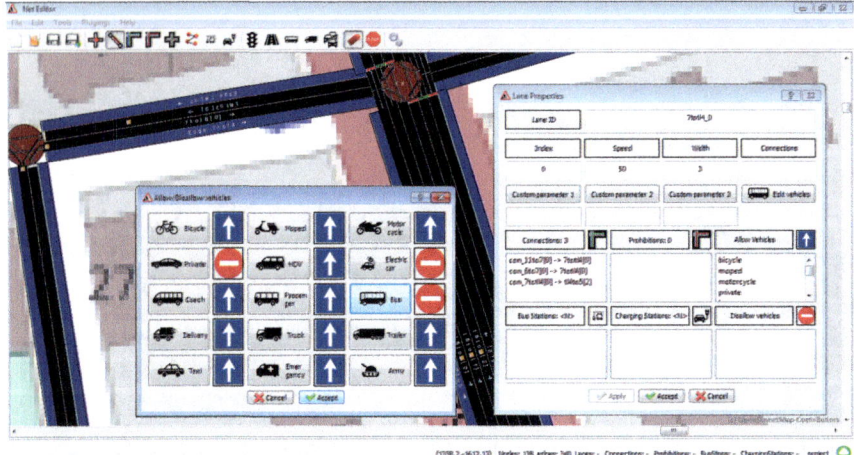

Fig. 6 Lane-based specifications' input dialog

The parameters *allow* and *disallow* refer to the abstract vehicle parameter *vehicle class*. For the intended functionality, vehicles need to be assigned the corresponding *vehicles class* parameter, as outlined later in Sect. 3.2.

If specified, lane-based specifications require a subsequent generation of the edge descriptions. Further notes regarding the order of generating individual network component definitions can be found later in Sect. 3.1.7. An example output is shown in Listing 3.4 that contains lane-based specifications in the edge definition file (typically .edg.xml).

Listing 3.4: Example edge definition with lane-specific parameters in the SUMO specified xml-format (typically .edg.xml)

```
<?xml version="1.0" encoding="utf-8"?>
<edges>
    <edge from="2" id="2to17" numLanes="2" numVeh="17050" priority="9" speed="
        13.889" to="17"/>
    <edge from="2" id="2to18" numLanes="2" numVeh="5400" priority="7" speed="
        13.889" to="18"/>
    <edge from="2" id="2to19" numLanes="2" numVeh="18600" priority="9" speed="
        13.889" to="19">
        <lane allow="bus, taxi" index="0" width="4"/>
    </edge>
    <edge from="3" id="3to2" numLanes="2" numVeh="4950" priority="7" speed="
        13.889" to="2"/>
    <edge from="3" id="3to4" numLanes="2" numVeh="4400" priority="7" speed="
        13.889" to="4"/>
    <edge from="3" id="3to5" numLanes="1" numVeh="975" priority="4" speed="8.333
        " to="5"/>
    ...
</edges>
```

3.1.4 Bus Stops

In order to model public transport behavior, recurring vehicle halts can be specified in SUMO using *busStop* objects. eNetEditor allows the definition of bus stops using an input dialog that can be called by the keyboard entry b and is shown in Fig. 7. The corresponding variable B consists of a list $B = (b_i)$ with…

$$\text{specification of bus stop } i \quad b_i, \tag{3.7}$$
$$\text{bus stop index} \quad i \quad = 1 \ldots b_{st} \text{ and}$$
$$\text{total number of specified bus stops} \quad b_{st}.$$

Each element b_i contains all parameters for the corresponding bus stop:

Listing 3.5: Example bus stop definition file in the SUMO specified xml-format

```xml
<?xml version="1.0" encoding="utf-8"?>
<additional>
  <busStop id="bS_2to19_0a" lane="2to19_0" startPos="10" endPos="25"/>
  <busStop id="bS_54to72_0a" lane="54to72_0" startPos="10" endPos="25"/>
  <busStop id="bS_111to113_0a" lane="111to113_0" startPos="10" endPos="25"/>
  <busStop id="bS_132to135_0a" lane="132to135_0" startPos="10" endPos="25"/>
  ...
</additional>
```

$$b_i = \begin{bmatrix} {}^B d_i \ l_i \ {}^B j \ {}^B x_{s,i} \ {}^B x_{e,i} \end{bmatrix}, \text{ with } \ldots \tag{3.8}$$

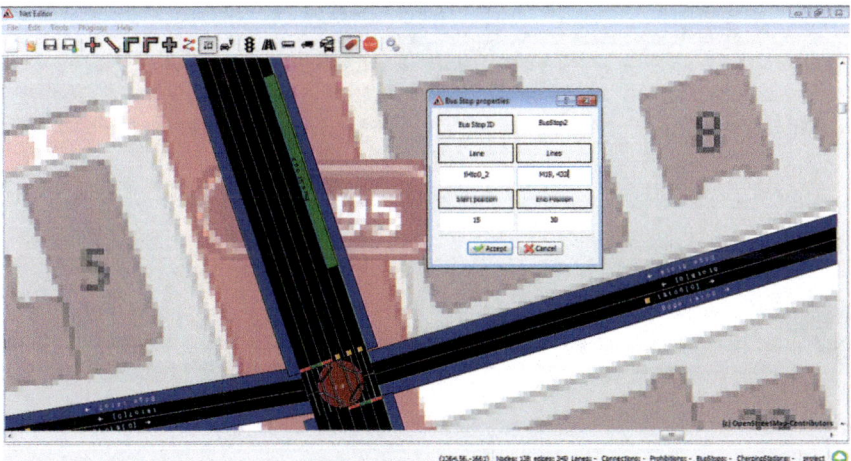

Fig. 7 Bus stop input dialog

$$\text{identifier of bus stop } i \quad {}^B d_i,$$
$$\text{lane on which bus stop } i \text{ is located} \quad l_i,$$
$$\text{index of bus stop } i \text{ on lane } l_i \quad {}^B j \quad = 1 \dots 26,$$
$$\text{start position of bus stop } i \text{ on lane } l_i \quad {}^B x_{s,i} \quad \text{(in m) and}$$
$$\text{end position of bus stop } i \text{ on lane } l_i \quad {}^B x_{e,i} \quad \text{(in m)}.$$

The output is a SUMO-compatible xml file, which is exemplified in Listing 3.5.

3.1.5 Charging Stations

Charging stations constitute the infrastructural complement to the energy consumption of vehicles. In analogy to bus stops, eNetEditor also allows the definition of charging stations, where compatible vehicles can be supplied with a specified power, if operational conditions allow. The placement and definition of bus stops in the network are similar to those of bus stops. The dialog for the definition of charging stations can be called in eNetEditor by the keyboard entry t. eNetEditor's dialog window is shown in Fig. 8. Variable T contains all of the network's charging station information in form of a list $T = (t_i)$ with...

$$\text{specification of charging station } i \quad t_i, \qquad (3.9)$$
$$\text{charging station index} \quad i \quad = 1 \dots t_{cs}, \text{ and}$$
$$\text{total number of specified charging stations} \quad t_{cs}.$$

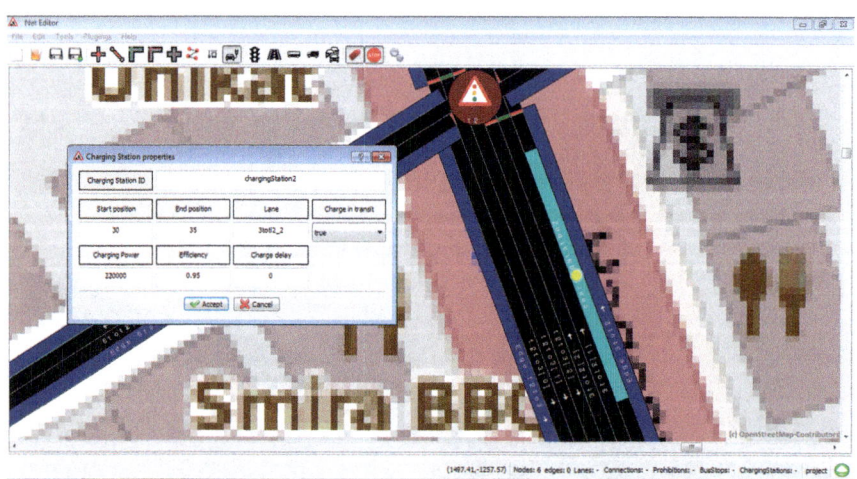

Fig. 8 Charging station input dialog

Each element t_i contains all parameters for the corresponding charging station:

Listing 3.6: Example bus stop definition file in the SUMO specified xml-format

```
<?xml version="1.0" encoding="utf-8"?> <additional>
  <chargingStation id="cS_2to19_0a" lane="2to19_0" startPos="10" endPos="25"
      chrgpower="200000" efficiency="0.95" chargeDelay="2" chargeInTransit="0"
      eType="a"/>
  <chargingStation id="cS_17to2_0a" lane="17to2_0" startPos="0" endPos="62"
      chrgpower="220000" efficiency="0.9" chargeDelay="5" chargeInTransit="1"
      eType="b"/>
  <chargingStation id="cS_17to2_1a" lane="17to2_1" startPos="0" endPos="62"
      chrgpower="250000" efficiency="0.9" chargeDelay="5" chargeInTransit="1"
      eType="a"/>
  ...
</additional>
```

$$t_i = \begin{bmatrix} {}^T d_i & l_i & {}^T j & {}^T x_{s,i} & {}^T x_{e,i} & {}^T P_i & {}^T \eta_i & T_i & d_i & c_i \end{bmatrix}, \text{with} \ldots \quad (3.10)$$

$$\text{identifier of charging station } i \quad {}^T d_i,$$
$$\text{lane on which charging station } i \text{ is located} \quad l_i,$$
$$\text{index of charging station } i \text{ on lane } l_i \quad {}^T j \quad = 1 \ldots 26,$$
$$\text{start position of charging station } i \text{ on lane } l_i \quad {}^T x_{s,i} \quad (\text{in m}),$$
$$\text{end position of charging station } i \text{ on lane } l_i \quad {}^T x_{e,i} \quad (\text{in m}),$$
$$\text{charging power of charging station } i \quad {}^T P_i \quad (\text{in W}),$$
$$\text{charging efficiency of charging station } i \quad {}^T \eta_i,$$
$$\text{delay between arrival and charging at charging station } i \quad T_i \quad (\text{in s})$$
$$\text{(dynamic) charging of moving vehicles above charging station } i \quad d_i \quad \text{and}$$
$$\text{compatible vehicle eType(s) that can be charged by } t_i \quad c_i.$$

The output is an xml file that is compatible with SUMO and the implementations are described in [6] for modeling the energy consumption of vehicles and their energy supply with the specified infrastructure elements. An output is exemplified in Listing 3.6.

3.1.6　Light Signal Systems

The definition of traffic light signal system plans for nodes of type *traffic_light* takes place during the network creation by netconvert. In most cases, the automatically generated programs/plans of a junction's light signal system will often deviate from its real behavior. SUMO allows multiple ways to interact with light signal systems, both during simulation definition and during runtime. Whereas traffic light signal systems can be declared as actuated for vehicle flow-based triggering or their program switched during simulation runtime using WAUT [9] or TraCI [10], it is also possible to declare static programs with phases of constant duration. Junction's programs can often be sufficiently approximated by a static program, if the period under observation is characterized by similar traffic intensities.

Each phase of a traffic light signal system in SUMO consists of a duration and a state, where the state is an aggregation of all connection/link states. While the indexing of links is (currently) carried out in clockwise order, one difficulty of defining programs beforehand is that link indices are defined by netconvert and are known only after the net file has been created. Therefore, the chosen solution for eNetEditor is to automatically create a net using netconvert and to parse for the lane connection order at the regarded junctions. Afterward, eNetEditor allows the definition and parameterization of light signal phases at the parsed junctions. Figure 9 exemplifies a traffic light signal-controlled junction ('40') in SUMO along with eNetEditor's input dialog and the generated text output.

3.1.7   Network Creation

After all constituents of the simulation's road network have been specified in eNetEditor's graphical user interface and user dialogs, the network can be created with the specified parameters using the sequence chart shown in Fig. 10. It is a multistage creation process that ultimately results in a call to netconvert, passing it all required files, to generate the net file (.net.xml). Bus stop and charging station definitions are created separately (in .add.bStop.xml and .add.chrg.xml, respectively) as *additional files* under the xml tag < additional >. The generated call to netconvert has the required syntax:

```
netconvert.exe -n genNets\ringSect3.nod.xml -e genNets\
ringSect3.edg.xml
  -o genNets\ringSect3.net.xml -x genNets\ringSect3.
con.xml
```

3.2   Vehicle Definition and Generation

The creation of a scenario by filling a defined network with life, i.e., vehicle objects with routes, initially requires defined vehicle types. SUMO allows the definition of various vehicle types along with their parameters that vehicle objects have access to. Most of these parameters are used by implemented vehicle models and include physical constraints and descriptions of the vehicle itself (e.g., maximum speed, vehicle length, color), driver-specific parameters (e.g., minimum gap between vehicles, impatience, deviation from speed limit) or model specifications (e.g., the car-following behavior, lane-change model, and other user-defined *devices*). Based on previous implementations described in [6], an *energy device* has been implemented

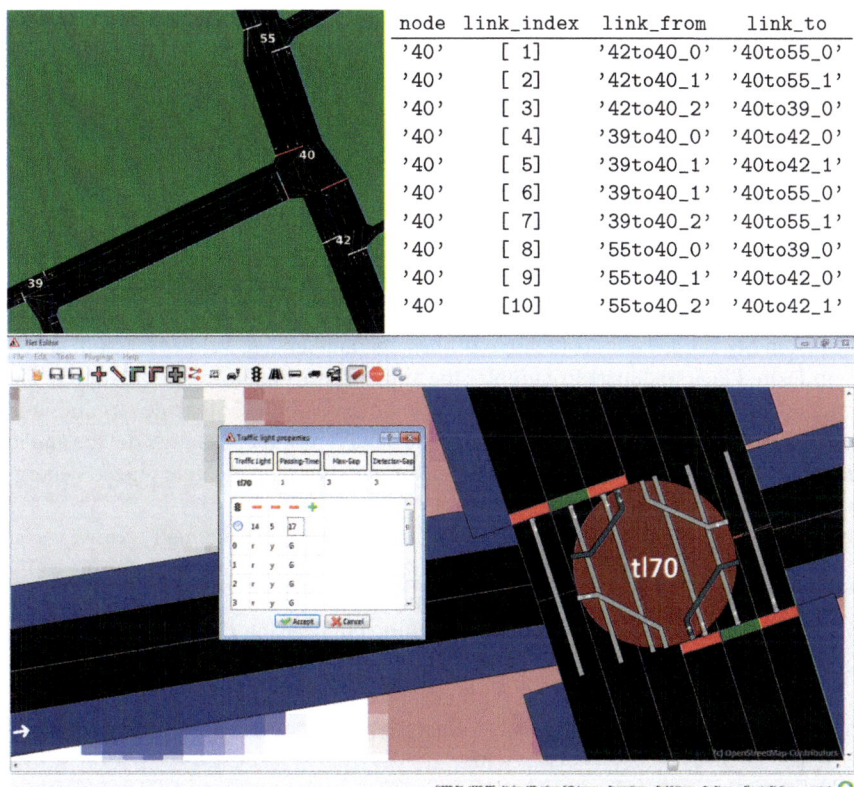

node	link_index	link_from	link_to
'40'	[1]	'42to40_0'	'40to55_0'
'40'	[2]	'42to40_1'	'40to55_1'
'40'	[3]	'42to40_2'	'40to39_0'
'40'	[4]	'39to40_0'	'40to42_0'
'40'	[5]	'39to40_1'	'40to42_1'
'40'	[6]	'39to40_1'	'40to55_0'
'40'	[7]	'39to40_2'	'40to55_1'
'40'	[8]	'55to40_0'	'40to39_0'
'40'	[9]	'55to40_1'	'40to42_0'
'40'	[10]	'55to40_2'	'40to42_1'

Fig. 9 Layout of traffic light signal-controlled junction '40' in SUMO (top left), output of its link indices for its program definitions (top right), and input dialog for definition of traffic light-controlled junctions in eNetEditor (bottom)

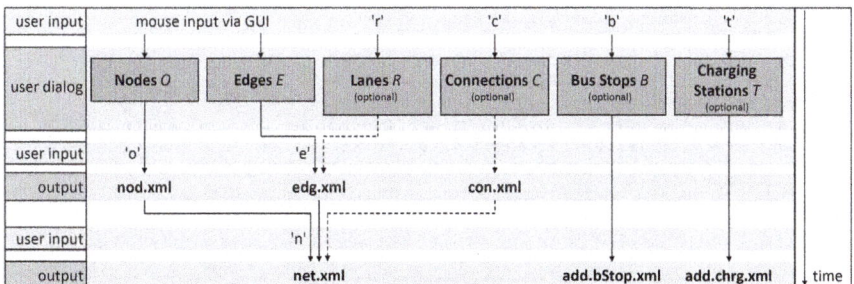

Fig. 10 Sequence chart of the network creation process

in SUMO that allows the realistic recreation of the energy variation of vehicles' energy content during simulation runtime, based on newly introduced and defined vehicle parameters, such as vehicle mass, maximum battery capacity (i.e., energy content), drivetrain efficiencies, or various drag coefficients. To attract a wide user base, the goal of these implementations was simplicity by introducing basic physical and comprehensible vehicle parameters.

After defining vehicle types, vehicle objects can be instantiated with *vehicle attributes* (initial values) for different model variables. These parameters are only contained within the vehicle objects themselves and are not part of the vehicle-type definition and therefore need to be set separately, e.g., manually or by a router described in Sect. 3.3. The most known of these parameters are vehicles' initial speed, their desired lane to depart from (enter the simulation), departure (simulation entry) time, or route. Figure 12 shows the structure of these vehicle attributes as definable in eNetEditor in form of a class diagram. The newly introduced energy device and its attributes are shown in green as well as the initial value of a vehicle object's energy content.

The configuration of vehicle attributes can be performed manually. However, in large simulation scenarios this task is carried out by the different routers during traffic demand generation and calibration that will be described in Sect. 3.3. In the following, an overview will be given on the data structure and user interface in eNetEditor for the definition and generation of vehicle types.

Vehicle types can be specified with all relevant parameters in eNetEditor using the user interface shown in Fig. 11. Vehicle-type properties are stored in the two-dimensional array $V^{n_v \times 27} = (v_i)$, with

Listing 3.7: Example vehicle type definition file in the SUMO specified xml-format with additional parameters used by SUMO's newly implemented energy device

```xml
<?xml version="1.0" encoding="utf-8"?>
<vType id="car" accel="2" decel="5" maxSpeed="42" length="4.5" minGap="2.5"
    sigma="0.5" speedDev="0.05" speedFactor="1.1" tau="1" impatience="0.1">
  <param key="MaxBatKap" value="20000"/>
  <param key="PowerMax" value="80000"/>
  <param key="Mass" value="2000"/>
  <param key="InternalMomentOfInertia" value="50"/>
  <param key="FrontSurfaceArea" value="2.0"/>
  <param key="AirDragCoefficient" value="0.4"/>
  <param key="RadialDragCoefficient" value="0"/>
  <param key="RollDragCoefficient" value="0.15"/>
  <param key="ConstantPowerIntake" value="40"/>
  <param key="PropulsionEfficiency" value="0.8"/>
  <param key="RecuperationEfficiency" value="0.5"/>
  <param key="EnType" value="a"/>
</vType>
```

$$\text{vehicle type index} \quad i \ = 1 \ldots n_v, \qquad (3.11)$$
$$\text{number of vehicle type definitions} \quad n_v.$$

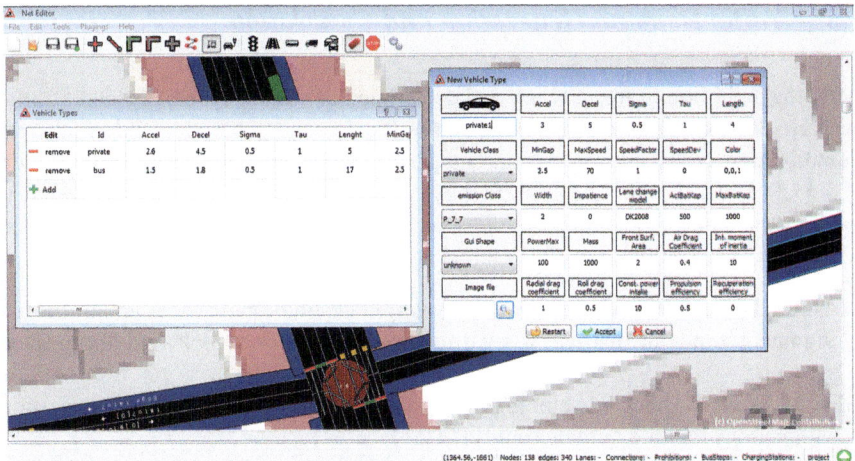

Fig. 11 Input dialog for the definition of vehicle types

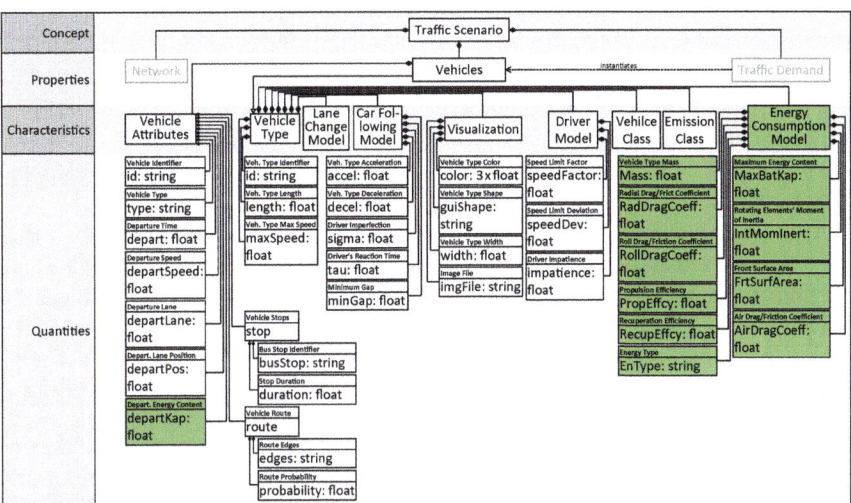

Fig. 12 Structure of vehicle and vehicle-type attributes as implemented in eNetEditor represented in form of a UML class diagram

In the same list-type format, as introduced in Sect. 3.1 for lanes, bus stops, and charging stations, each element v_i contains 28 vehicle-type parameters that are exemplified in Listing 3.7. It also shows the required syntax for SUMO along with the implemented energy device (Fig. 12).

3.3 Traffic Demand Generation and Calibration

After the definition of the network and declaration of vehicle types, the target traffic scenario can be finalized by a subsequent traffic demand generation and, if desired, its calibration. Initial and most important part of vehicle instantiation is the definition of vehicle routes. For this purpose, SUMO comes with a variety of routers. Secondly, initial parameters of the routed vehicle objects need to be set. SUMO's mostly applied routers for the creation of traffic demand are

MAROUTER: Macroscopic data from origin/destination matrices and traffic assignment zones or districts is used to assign routes to vehicles within a given network.

DFROUTER: Detected vehicle flow data is used to calculate the flow proportion at junctions to build vehicle routes [11].

JTRROUTER: Definitions of traffic flows on edges and turn ratios at junctions are used to compute vehicle routes within a given network.

In order to utilize MAROUTER, *SUMO Traffic Modeler* has been developed [12], which is a very functional third-party tool that supports the generation traffic demand within an existing network. It aims at supplementing a network with demographic data to create time-varying origin/destination data, which can be used by SUMO's MAROUTER for the generation of vehicle routes. Therefore, eNetEditor focuses on utilizing SUMO's JTRROUTER and (more importantly) DFROUTER for the rapid generation of traffic demand, which is described in more detail in Sect. 3.3.1.

After an initial set of desired routes (edge-by-edge declarations) or trips (origin/destination declarations) has been generated by a router that most often does not take into account the resulting interaction between vehicles, the route choices for individual vehicles need to be calibrated to create realistic traffic scenarios by distributing vehicles iteratively among alternative routes within the network. Two of the most widely utilized applications for traffic calibration are

DUAROUTER: Based on [13], this user assignment algorithm aims at finding one dynamic user equilibrium for given vehicle trips by assigning routes iteratively until no vehicle can reduce its travel time by an alternative route choice.

cadyts: Based on [14], this method further takes into account a supply model of the network to iteratively find a stationary condition, which is consistent with existing traffic counts.

Figure 13 shows the implemented data structure for generating and calibrating traffic demand that the following sections will give more details on. The necessary steps in eNetEditor will further be outlined to build calibrated traffic scenarios. The components shown in red are not implemented in eNetEditor; for creating routes based on origin/destination matrices with MAROUTER, *SUMO Traffic Modeler* (also referred to as *SUMO Traffic Generator*) [12] should be used instead.

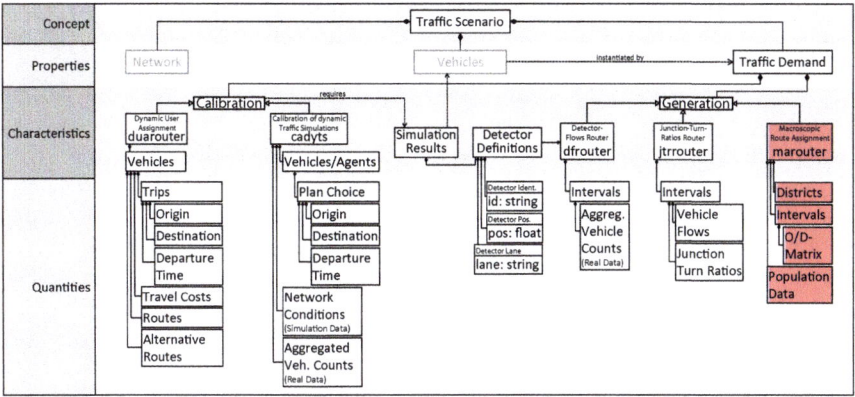

Fig. 13 Class diagram of traffic demand generation and calibration structure in SUMO and eNetEditor

3.3.1 Traffic Demand Generation

For the generation of vehicle instances and routes, eNetEditor offers two possibilities.

JTRROUTER

For demand generation with JTRROUTER, eNetEditor allows the definition of vehicle flows and junction turn ratios.

Flow definitions are stored in the two-dimensional array $F^{n \times 8} = (f_{i,k})$, where

$$\text{flow index} \quad i \quad = 1 \ldots n, \qquad (3.12)$$
$$\text{property index} \quad k \quad = 1 \ldots 8 \text{ and}$$
$$\text{number of flow definitions} \quad n.$$

Each element f_i contains all required parameters of a flow definition:

$$f_i = \begin{bmatrix} {}^F d_i & {}^F t_{s,i} & {}^F t_{e,i} & {}^F e_i & {}^F V_i & {}^F T_i & {}^F l_i & {}^F v_i \end{bmatrix}, \text{with} \ldots \qquad (3.13)$$

$$\text{identifier of flow definition } f_i \quad {}^F d_i,$$
$$\text{start time of flow definition } f_i \quad {}^F t_{s,i},$$
$$\text{end time of flow definition } f_i \quad {}^F t_{e,i},$$
$$\text{origin edge of flow definition } f_i \quad {}^F e_i,$$
$$\text{number of vehicles in flow definition } f_i \quad {}^F V_i,$$
$$\text{vehicle type of flow definition } f_i \quad {}^F T_i,$$
$$\text{departure (initial) lane of flow definition } f_i\text{'s vehicles} \quad {}^F l_i \quad \text{and}$$
$$\text{departure (initial) speed of flow definition } f_i\text{'s vehicles} \quad {}^F v_i.$$

The resulting xml output is shown in Listing 3.8.

Listing 3.8: Example vehicle flow definition file in the SUMO specified xml-format

```xml
<?xml version="1.0" encoding="utf-8"?>
<flows>
    <flow id="86to11flow1" begin="0" end="8640" from="86to11" number="88" type="
        car" departLane="random" departSpeed="random"/>
    <flow id="87to13flow1" begin="0" end="8640" from="87to13" number="90" type="
        car" departLane="random" departSpeed="random"/>
    <flow id="88to13flow2" begin="0" end="8640" from="88to13" number="15" type="
        bus" departLane="random" departSpeed="random"/>
    ...
</flows>
```

The typical use case of JTRROUTER is that incoming flows into a network are defined at the network boundaries, along with turn ratio definitions at junctions to fill the network with vehicle interaction. When the vehicle flow input dialog is opened for the first time, vehicle flows are initialized for all incoming edges into the regarded network, whose parameters have to be specified. The set of incoming edges E_I is defined by Eq. 3.14.

$$E_I := \left\{ e_{i,j,1} \left| \forall j \in [0, n] : e_{i,j,1} = 0 \right. \right\}. \tag{3.14}$$

In addition to vehicle flows, turning ratios need to be defined at junctions to feed JTRROUTER with all required information. The concept behind the structure for the turn ratio definitions at junctions is similar to that of connections described in Sect. 3.1.2. The main difference between the two declarations is that connections are lane-based definitions (each connection is described by an incoming and outgoing *lane*), where turning ratios are edge based (each turn ratio is described by an incoming and outgoing *edge*). Consequently, the complexity for turn ratio definitions at junctions is lower than that of connections and can be reduced to three degrees of freedom: the incoming and outgoing edges, each described by their origin and destination nodes, where the destination node of the incoming edge is identical to the origin node of the outgoing edge. Each resulting turn ratio is a sequence of three nodes, attributed with a probability. The chosen data structure for turn ratio definitions is a three-dimensional matrix implemented as $J^{n \times n \times n}$, with

$$
\begin{aligned}
\text{origin node index of incoming edge } k &= 1 \dots n, \quad (3.15) \\
\text{destination/origin node index of incoming/outgoing edge } l &= 1 \dots n, \\
\text{destination node index of outgoing edge } m &= 1 \dots n, \\
\text{number of nodes } n & \quad \text{and} \\
\text{turn ratio (probability) } j_{k,l,m} &= 0 \dots 1.
\end{aligned}
$$

Even if the size of J grows exponentially with the amount of nodes, the resulting variable is a sparse matrix, whose required amount of memory can correspondingly be reduced by explicitly declaring it as such (e.g., in MATLAB with the command *sparse*). Since junction turn ratio definitions are treated as probabilities when JTR-ROUTER creates vehicle routes, the definitions have to meet requirements to be valid. For the conservation of vehicle flows at junctions, Eq. 3.16 defines that the turning ratios from an incoming edge (taken from the adjacency matrix $E = (e)$) must add up to 1. The resulting variable J is automatically checked accordingly for validity when entering turning ratios. Listing 3.9 exemplifies the xml output for the definition of junction turn ratios.

Listing 3.9: Example junction turn ratio definition file in the SUMO specified xml-format

```
<?xml version="1.0" encoding="utf-8"?>
<turns>
   <fromEdge id="23to29">
      <toEdge id="29to30" probability="0.2"/>
      <toEdge id="29to33" probability="0.3"/>
      <toEdge id="29to34" probability="0.5"/>
   </fromEdge>
   ...
</turns>
```

$$\sum_m j_{k,l,m} = \begin{cases} 1 & \text{if } e_{k,l,1} \neq 0 \\ 0 & \text{else.} \end{cases} \tag{3.16}$$

The vehicle flow and junction turn ratio input dialogs are shown in Fig. 14 and can be called by the keyboard entries of f and c, respectively (due to their similarity, junction turn ratios can be defined along with connections in the lower half of the same input window as shown in Fig. 5).

If the vehicle flow and junction turn ratio definitions have been created, the specified scenario can be built with a keyboard entry j. This (1) calls JTRROUTER with all required options, which returns vehicle routes that (2) are used to build the resulting SUMO configuration (scenario) file shown in Listing 3.10. The following syntax exemplifies the automated execution of JTRROUTER with all required options:

```
jtrrouter --random 1 -f scenario.flows.xml -t scenario.
turns.xml
   -n scenario.net.xml -o scenario.rou.xml
```

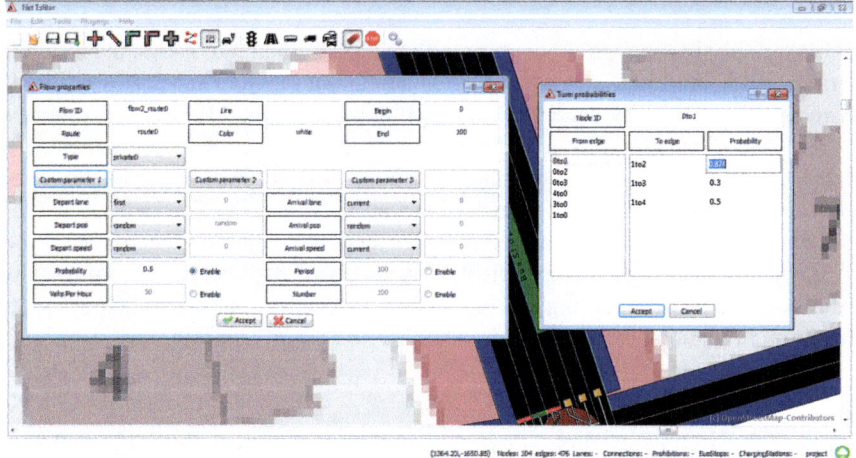

Fig. 14 Input dialog for vehicle flow (left) and junction turn ratio (right) definitions

Listing 3.10: Generic SUMO configuration file as built by eNetEditor using vehicle route definitions in `scenario.rou.xml`, either created manually or by any of SUMO's routers

```
<configuration>
  <input>
    <net-file value="scenario.net.xml"/>
    <additional-files value="scenario.add.chrg.xml,scenario.add.bStop.xml"/>
    <route-files value="scenario.rou.xml"/>
  </input>
  <output>
    <tripinfo-output value="scenario.tripinfo.out.xml"/>
    <battery-output value="scenario.battery.out.xml"/>
  </output>
</configuration>
```

DFROUTER

If the regarded road network is equipped with induction loops or other sensors to measure vehicle traverses, the measurement data can be used to build vehicle routes using DFROUTER, which performs a stochastic route estimation for vehicles. eNetEditor offers an interface to DFROUTER that will be described in this section.

For the definition of induction loops (detectors), the keyword *numVehs* has been reserved as a property name for edges described in Sect. 3.1.1. If vehicle counts exist for an edge, their values can be assigned to the edge attribute *numVehs* for subsequent processing to build a valid call to DFROUTER, including detector and vehicle count definitions. DFROUTER declares routes for two vehicle types: *PKW* and *LKW*, representing the average passenger and commercial/freight vehicle in the simulation scenario, respectively. PKW and LKW need to be declared as vehicle types in eNetEditor before building the scenario using DFROUTER's routes.

Listing 3.11: Example induction loop (detector) definition used as an input in the call to DFROUTER

```xml
<?xml version="1.0" encoding="utf-8"?>
<detectors>
   <detectorDefinition id="det0_1to5_0" lane="1to5_0" pos="0"/>
   <detectorDefinition id="det1_1to11_0" lane="1to11_0" pos="0"/>
   <detectorDefinition id="det2_1to13_0" lane="1to13_0" pos="0"/>
   <detectorDefinition id="det3_1to13_1" lane="1to13_1" pos="0"/>
   <detectorDefinition id="det4_1to17_0" lane="1to17_0" pos="0"/>
   ...
</detectors>
```

Listing 3.12: Example of the synthetically generated vehicle flow measurement file used as an input in the call to DFROUTER

```
Detector;Time;qPKW;vPKW;qLKW;vLKW
det0_1to5_0;0;4.2989;7;0.22626;5
det0_1to5_0;20;8.334;7;0.43863;5
det0_1to5_0;40;10.5241;7;0.5539;5
det0_1to5_0;60;16.2798;7;0.85683;5
det1_1to11_0;0;4.4204;7;0.23265;5
det1_1to11_0;20;7.1806;7;0.37793;5
det1_1to11_0;40;12.4429;7;0.65489;5
det1_1to11_0;60;16.4115;7;0.86376;5
det2_1to13_0;0;34.4494;7;1.8131;5
det2_1to13_0;20;67.7355;7;3.565;5
det2_1to13_0;40;94.3726;7;4.967;5
det2_1to13_0;60;109.7726;7;5.7775;5
det3_1to13_1;0;31.8458;7;1.6761;5
det3_1to13_1;20;63.9636;7;3.3665;5
det3_1to13_1;40;85.5032;7;4.5002;5
det3_1to13_1;60;125.9959;7;6.6314;5
...
```

Detector Definitions: Detector definitions are built according to the declaration of the attribute *numVehs* on the corresponding edge, as exemplified in Listing 3.11.

Vehicle Counts: Synthetic vehicle counts are generated according to the specified value of the *numVehs* attribute. Their time stamps are distributed by default according to the daily traffic curve *TGw2West* from [15] (Figs. 2, 3 and 4) or by an arbitrary user-specifiable distribution function, within a user-specified time period (e.g., 09:00–12:00) with the user-specified aggregation periods (e.g., 20 min). In a last step, the user defines the distribution of passenger and commercial/freight or, by default, a constant distribution of 95% passenger and 5% commercial/freight vehicles is assumed. The generated vehicle flow measurement file is exemplified in Listing 3.12.

If all required definitions have been made, the generation of a scenario with vehicle routes from DFROUTER can be initiated by the keyboard entry *d*. Much like the process defined in Sect. 3.3.1 for JTRROUTER, this (1) calls DFROUTER with all required options. DFROUTER then returns two files: a route file containing a list of all generated routes and a vehicle instantiation file (with reference to the declared routes). These files (2) are used to build the resulting SUMO configuration, which is comparable to Listing 3.10. The following command exemplifies the DFROUTER execution to build the described routes:

```
dfrouter --random 1 --net-file scenario.net.xml
-detector-files
   scenario.det.xml --measure-files scenario.flowMeas.
xml --routes-
   output scenario.rou.xml --emitters-output scenario.
veh.xml --time-
   step 1200 --departlane random --departspeed random
-vtype true
```

Future work on eNetEditor for a more realistic integration and application of DFROUTER's functionalities includes user-specifiable lane-based distributions of vehicles.

The sequence chart for traffic demand generation with JTRROUTER and DFROUTER is shown in Fig. 15.

3.3.2 Traffic Demand Calibration

The problem of the routers described in Sect. 3.3.1 is that vehicle instances are assigned a route under static conditions, e.g., by searching for the shortest path through the network. Prevailing traffic conditions are not regarded. This task can be compared to that of finding a/the short/est way through a traffic-intense urban area, only by knowledge about the nodes and edges in its network, in hope that the chosen route will be the fastest or the one with the least amount of energy required. It is obvious that prevailing traffic conditions can have a high impact on these optimization criteria. In real traffic, participants without (sometimes also with) assistance usually need a few tries, where the iterative variation of departure time and/or route choice ultimately yields the *optimal* perceived route for a recurring trip.

A variety of methods exist for traffic simulations that aim to minimize a cost function to find an optimum route distribution among participants (vehicles) in a similar manner as described above. In the scope of this work, *demand calibration* refers to the adaption of these route assignments. A comparison between different traffic demand assignment methods can be found in [16]. eNetEditor provides an interface to two established implementations for SUMO that aim to optimize the route assignment for vehicle instances: DUAROUTER and cadyts. The basis for both is a validated set of vehicle trips, each consisting of origin, destination, and departure

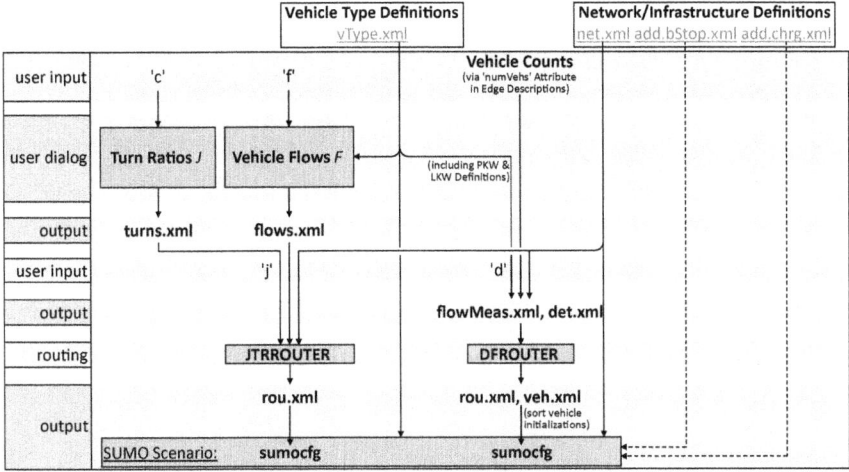

Fig. 15 Sequence chart of the traffic demand creation process using JTRROUTER and/or DFROUTER

time. These parameters are taken from the route definitions that were stochastically determined by DFROUTER.

The calibration of traffic demand in eNetEditor is implemented as a two-phase process:

1. DUAROUTER: Vehicle routes are distributed by creating alternative routes and evaluating the resulting edge-based weights for specified vehicle trips, until no vehicle can modify its route choice without increasing its travel cost. One of the outputs is a vehicle route file for vehicles containing alternative route probabilities and corresponding travel costs.
2. cadyts: cadyts itself performs no routing. The alternative routes created by DUAROUTER are evaluated and vehicle route choices adapted, such that the chosen routes comply with detector measurements passed as an input to cadyts.

If traffic scenarios are to be calibrated with cadyts, a previous calibration with DUAROUTER is required in eNetEditor.

DUAROUTER

While DUAROUTER itself has been implemented in C++, the iterative process of finding alternative routes and executing the simulation has been implemented in form of a Python script, *dualterate.py*. eNetEditor automates the creation and execution of the required command-line argument by the keyboard entry of the button *u*, if previous vehicle routes from DFROUTER (3.3.1) are present. These routes are stripped down to each of their trip definitions using the existing Python script *route2trips.py*: the route's origin edge, destination edge, and departure time. A trip definition is exemplified in Listing 3.13.

Listing 3.13: Example trip definition as a result of route definitions from DFROUTER used as an input in the call to DUAROUTER

```xml
<?xml version="1.0" encoding="utf-8"?>
<trips>
  <trip depart="0.00" departLane="random" departPos="0" departSpeed="random"
      from="82to49" id="emitter_det211_82to49_0_0" to="13to16" type="PKW"/>
  <trip depart="0.00" departLane="random" departPos="0" departSpeed="random"
      from="83to33" id="emitter_det212_83to33_0_0" to="10to27" type="PKW"/>
  <trip depart="0.00" departLane="random" departPos="0" departSpeed="random"
      from="84to23" id="emitter_det213_84to23_0_0" to="13to16" type="PKW"/>
  <trip depart="0.00" departLane="random" departPos="0" departSpeed="random"
      from="85to10" id="emitter_det214_85to10_0_0" to="23to28" type="PKW"/>
  <trip depart="0.00" departLane="random" departPos="0" departSpeed="random"
      from="85to10" id="emitter_det215_85to10_1_0" to="4to9" type="PKW"/>
  <trip depart="0.00" departLane="random" departPos="0" departSpeed="random"
      from="86to11" id="emitter_det216_86to11_0_0" to="13to16" type="PKW"/>
  ...
</trips>
```

The resulting output in eNetEditor is a command-line argument that executes the iterative DUAROUTER Python script with the specified options as exemplified in the following command:

```
duaIterate.py -n scenario.net.xml -t scenario.trips.xml
-x detailed
 -+ "scenario.vtypes.xml,scenario.add.chrg.xml,scenario.
add.bStop.xml"
 -l 20
```

cadyts

If DUAROUTER has finished execution, a subsequent call to the Python script *cadytsIterate.py* can be made from eNetEditor by keyboard entry of the button *y*. cadyts requires traffic flow measurements in the xml format exemplified in Listing 3.14. In regard to the contained information, this measurement file is identical to that required by DFROUTER (with an additional user-specifiable standard deviation for each measurement), only in a different format. Currently, only one percentage value will be queried by eNetEditor, which derives the standard deviation as a fixed percentage of the corresponding flow for all measurement values. The measurement file is constructed from the values of the user-specified edge parameter *numVehs* in the same manner as described in Sect. 3.3.1 for DUAROUTER.

The resulting output is a command-line argument that executes cadyts with the specified options to iteratively adapt the route choice of vehicles. The following exemplifies a command to cadyts:

```
cadytsIterate.py -b 0 -n scenario.net.xml -d scenario.
calibFlowMeas.xml
 -+ "scenario.vtypes.xml,scenario.add.chrg.xml,
scenario.add.bStop.xml"
```

Listing 3.14: Example flow measurement definition created for subsequent call to DUAROUTER

```xml
<?xml version="1.0" encoding="utf-8"?>
<measurements>
   <onlink start="0"     end="1199" link="1to5"  value="4.280"  stddev="0.428"
      type="COUNT_VEH"/>
   <onlink start="1200" end="2399" link="1to5"  value="8.158"  stddev="0.816"
      type="COUNT_VEH"/>
   <onlink start="2400" end="3599" link="1to5"  value="12.036" stddev="1.204"
      type="COUNT_VEH"/>
   <onlink start="3600" end="4799" link="1to5"  value="15.914" stddev="1.592"
      type="COUNT_VEH"/>
   <onlink start="4800" end="5999" link="1to5"  value="19.390" stddev="1.939"
      type="COUNT_VEH"/>
   <onlink start="0"     end="1199" link="1to11" value="4.280"  stddev="0.428"
      type="COUNT_VEH"/>
   <onlink start="1200" end="2399" link="1to11" value="8.158"  stddev="0.816"
      type="COUNT_VEH"/>
   <onlink start="2400" end="3599" link="1to11" value="12.036" stddev="1.204"
      type="COUNT_VEH"/>
   ...
</measurements>
```

```
-r scenario.rou.alt.xml -W scenario.flowsEvaluation.
txt -l 20 -a 1200
```

3.4 Modules and Extensibility

This section shall give an overview on the modular structure of eNetEditor. The current module implementations allow users to extend eNetEditor's functionality arbitrarily, by defining and programming custom modules. A C++ template for *Qt Creator* will be made available for this purpose. Once compiled, the resulting Dynamic Link Library (DLL) can be placed into eNetEditor's directory under ...\plugins\, prompting eNetEditor to load it at the program start and dynamically integrating it into the interface. Modules are divided into four types that will be explained in the following sections.

3.4.1 Network Modification

The elementary module of eNetEditor contains all required routines that are responsible for modifying elementary objects of an existent scenario (nodes, edges, connections, vehicles, etc.). Their implementation allows users to add new arbitrary parameters and values for any existing object. While eNetEditor's database is implemented to store all user-specified information, users need to ensure that these additional parameters get processed, when passed to subsequent tools, such as netconvert or SUMO itself.

3.4.2   Including New Objects

The evaluation of simulations often requires new objects (implemented in SUMO or external tools) to obtain relevant data (e.g., certain vehicle types or new objects with user-specifiable properties/parameters). Modules are required that allow the representation of these new objects in eNetEditor's and their inclusion into the database. Modules will be provided with a set of functions to declare and edit new objects with custom data types and parameters in eNetEditor, comparable to interfaces used for the declaration of standard objects and their parameters, such as nodes, edges, or vehicles. These modules are aimed to provide a graphic user interface for and to work with custom modifications within SUMO. An example would be the introduced energy device in SUMO and an interface for eNetEditor for the input of newly introduced vehicle parameters.

3.4.3   Data Processing

Generally, obtained simulation data needs to be analyzed subsequently. This task usually consumes much time, where users apply a variety of different tools for the implementation of custom parsers and/or visualization routines. It further leaves the users with the problem of having to deal with large files. Data processing modules provide a set of functions to obtain and process data contained within xml-formatted output files. Plotting functions will be made available using the *QCustomPlot* library, with the idea of simplifying this task.

3.4.4   Integration of External Tools

The process of creating a simulation for SUMO consists of network definition and traffic inclusion. Due to its open-source character and subsequent wide user range, many external tools have been developed and contributed to SUMO, such as for vehicle routing and traffic calibration (e.g., DFROUTER, DUAROUTER, cadyts). An additional type of modules shall be responsible to call or execute external tools with user-specifiable options. The main advantage is the correct use of these extensions, which usually come with very little documentation on their correct usage and are therefore prone to be used incorrectly. Modules have been developed for creating traffic by generating calls to DFROUTER and JTRROUTER as well as for calibrating scenarios with DUAROUTER and cadyts.

3.4.5   Database Modification Modules

For the evaluation of certain scenarios, it would be most convenient to provide users a method of performing automated, script-based modifications to a scenario. An example would be the analysis of allowing electric vehicles to travel on bus lanes. In larger networks, performing these changes would take very long and be very prone

Listing 3.15: Example script that modifies the parameter *allow* of all bus lanes in a network

```
VAR sharedLanes = GET ALL lane WHERE lane.allow == 'bus'
SET sharedLanes.allow TO 'bus, electricVehicle'
```

to forgetting a bus lane. For this purpose, database modification modules provide a simple set of commands for users to perform automated changes within a scenario. The idea is to implement an interpreter and to provide simple commands, similar to SQL, to interface with the existing database. For the above example, Listing 3.15 exemplifies the script for changing the *allow* property of all lanes that allow buses to further allow a new vehicle class.

3.5 User Interaction and Interface

Beyond the described user dialogs, user interaction is implemented using hot keys. eNetEditor can be executed with the following command-line argument from MATLAB:

```
netBuild('quadraticBackgroundImage.png',widthInMeters,
'projectName')
```

The first command argument specifies the filename of a (quadratic) background image for visualization, representing the network to be examined. The second argument specifies the width (= height) in meters as an integer of the regarded network to be modeled. The third argument specifies the project name as a string.

Projects can be saved in the main window at any time with the keyboard entry *s*. If an existing project shall be opened, the above command-line argument must be executed in MATLAB with its project name specified in the third argument. After all windows have finished loading, previously saved projects can be loaded with the keyboard entry *l*. An overview of all hot keys can be found at the bottom of the main window.

4 Example Scenario and Results

An example scenario was created that represents a section of Braunschweig's road traffic. The scenario originates from the project *emil* (Elektromobilität mittels induktiver Ladung—electric mobility via inductive charging), which has integrated an inductive vehicle charging system and a compatible prototype bus fleet into Braunschweig's traffic and public transport infrastructure. Funded by the German Federal Ministry for Transport and Digital Infrastructure, goals of project *emil* include the analysis of the possible integration of private motor vehicles for the shared usage of the installed charging stations and other road infrastructure. With the application of the implementations presented in [6], various future traffic and scaling (market integration) scenarios will be simulated with SUMO to analyze and evaluate different aspects of the energy consumption.

In a first step, traffic measurements were taken from the city's traffic intensity map [17] and used for calibration as described in Sect. 3.3. In addition, one bus of the local public transport operator Braunschweiger Verkehrs-GmbH was equipped with a data logger, recording the GPS-based position, basic drivetrain, and data relevant for fleet management over the standardized Fleet Management System Interface [18]. The bus mainly operated on the designated line to be electrified, but essentially made several trips on most lines during the measurement period of a year. Relevant operational parameters have been extracted and will be used to model and parameterize public transport behavior of the corresponding vehicle instances in SUMO on the one hand and to validate the resulting traffic scenario on the other.

Figure 16 depicts eNetEditor's user interface along with the resulting scenario (left) next to its visualization in SUMO's graphical user interface. The SUMO corresponding configuration file, using the vehicle energy device and charging stations, after calibration with cadyts is shown in Listing 4.1.

The resulting output file (in this case `iteration_049.battery.out.xml`) can be used by subsequent analyses and optimization algorithms to determine (1) where in the network energy is consumed and (2) which locations are suitable for the placement of charging stations to supply vehicles with adequate energy *on the fly*, i.e., during their operation. Figure 17 shows first results that aim to supplement these future analyses.

Whereas the left image shows the cumulated energy that was consumed by all vehicles in the simulation, the center and right images focus on the energy consumption's complement: the time, which can be used for transferring energy into vehicles. If vehicles can be charged while moving at arbitrary speeds (e.g., catenary/trolley systems), any location within the road network can potentially be used for transferring energy into vehicles. The resulting cumulated time of all vehicles in the example scenario as a function of vehicle position is shown in the center image ($T_{st} \geq 0$ s). However, since most charging stations (e.g., gas stations) require vehicles to be standing still for a certain duration, halts can only be used for recharging if they have a minimum duration. The amount of energy that could potentially be used for charging reduces. The image to the right shows the cumulated duration of vehicle halts, of which each exceeded 10 s in the scenario as a function over vehicle position.

5 Charging Station Optimization

With the goal of optimizing charging station locations to provide a maximum amount of vehicles with adequate energy for their chosen routes, existing algorithms have

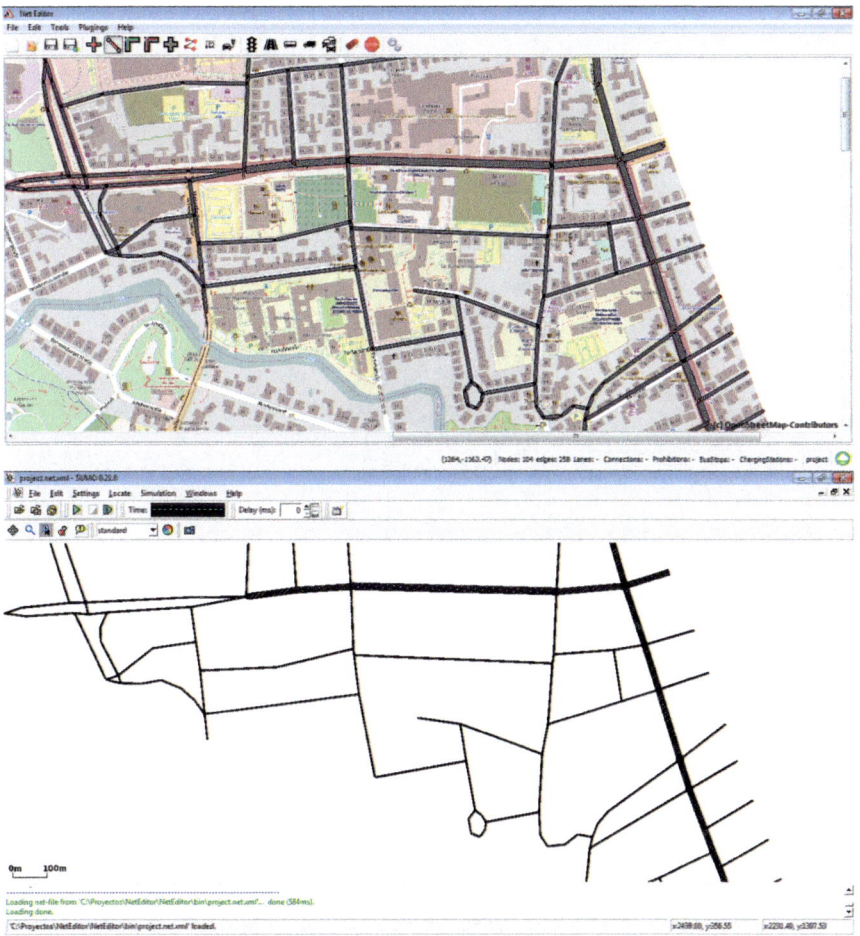

Fig. 16 eNetEditor's graphical user interface showing a user-specified background map and the corresponding network's digraph (left) and a snapshot of the resulting scenario in SUMO's graphical user interface (right) (c) OpenStreetMap contributors

been studied and adapted, which are referred to in the literature as *flow capture* [19], *discretionary service facility* [20], and *flow refueling* [21] optimization problems.

5.1 Background

The traffic is modeled in [19–21] by using graphs, consisting of edges and nodes, where intersections are represented by nodes and streets (lanes) by edges. Traffic is modeled as weights assigned to each edge that stand for the flow of vehicles along

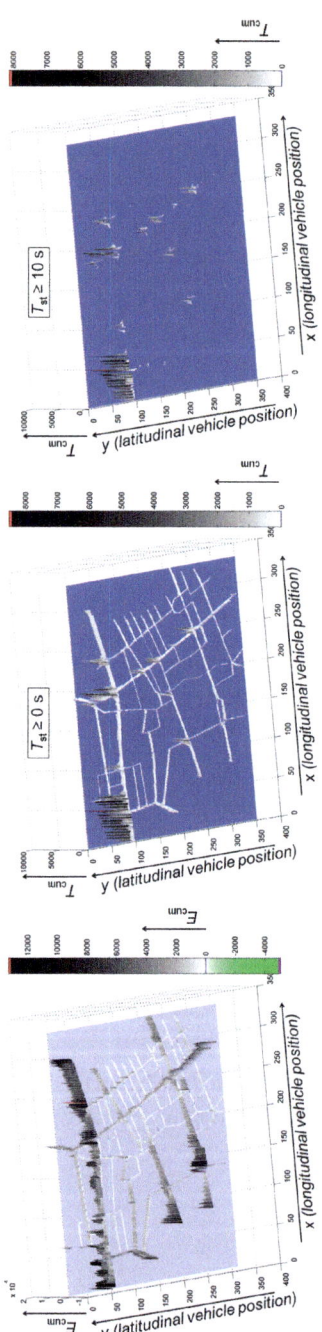

Fig. 17 Cumulated energy consumed (left) and cumulated vehicle standstill duration T_{st}, with $T_{st} \geq 0$ s (left) and $T_{st} \geq 10$ s (right), of all vehicles in the simulation over their position

Listing 4.1: SUMO configuration of a scenario calibrated with cadyts

```xml
<?xml version="1.0" encoding="utf-8"?>
<configuration ...>
  <input>
    <net-file value="ringSect5.net.xml"/>
    <route-files value="ringSect5_049.cal.xml"/>
    <additional-files value="dua_dump_049.add.xml,ringSect5.DFvtypes.xml,
      ringSect5.add.chrg.xml,ringSect5.add.bStop.xml"/>
  </input>
  <output>
    <battery-output value="iteration_049.battery.out.xml"/>
    <battery-output.precision value="4"/>
    ...
  </output>
  ...
</configuration>
```

the edge over a predefined period of time. The goal of these models is to maximize the flow that can be *captured* by the *discretionary service facilities* that are placed in the network. These models are suitable for regarding networks, where the stop of a traffic participant does not jam any followers (e.g., pedestrians on sidewalks) and where the utilization of a service facility does not take any time or the time spent is an irrelevant factor (e.g., for radar traps or for commercial placement).

The common property of each facility is that if a vehicle passes one of these, it is regarded as *captured*. The purpose of the service facility is fully fulfilled, and the implemented optimization algorithm will have no interest in that vehicle passing another facility. This assumption does not hold, if these facilities are interpreted as charging, gas, or refueling stations for vehicles. Vehicles cannot fully charge by simply passing a charging station. Instead, they require a minimum amount of time $t_{s,\mathrm{min}}$ for the charge to be initiated and after $t_{s,\mathrm{min}}$ the energy charged is proportional to the remaining time t_{charge} that the vehicle stays at the service facility, i.e., charging station. Also, more than one charge may be required by a vehicle along its route. This dependency needs to be regarded in the further development of these models, especially in the case of introducing electric vehicles, where the energy transfer rate (charging power) is only a small fraction of that of gasoline stations.

The models in [19–21] further assume that service facilities that are placed on a node capture the flows of each of that node's incoming/outgoing edge; thus, the search for optimal locations can be reduced to nodes only. This assumption can be applied, if the regarded network is represented in such a detail, that not only junctions are modeled as nodes. The provided examples in [19, 21], however, prove otherwise, as shown in Fig. 18. Road traffic vehicles are obliged in most countries (by law) not to stop on junctions, which clearly rules out junctions as suitable locations for charging stations because vehicles are supposed to just pass through.

The required modification of these models is formulated and explained in Sect. 5.2 for taking into account the energy that can be charged into vehicles located within the required proximity of a charging station.

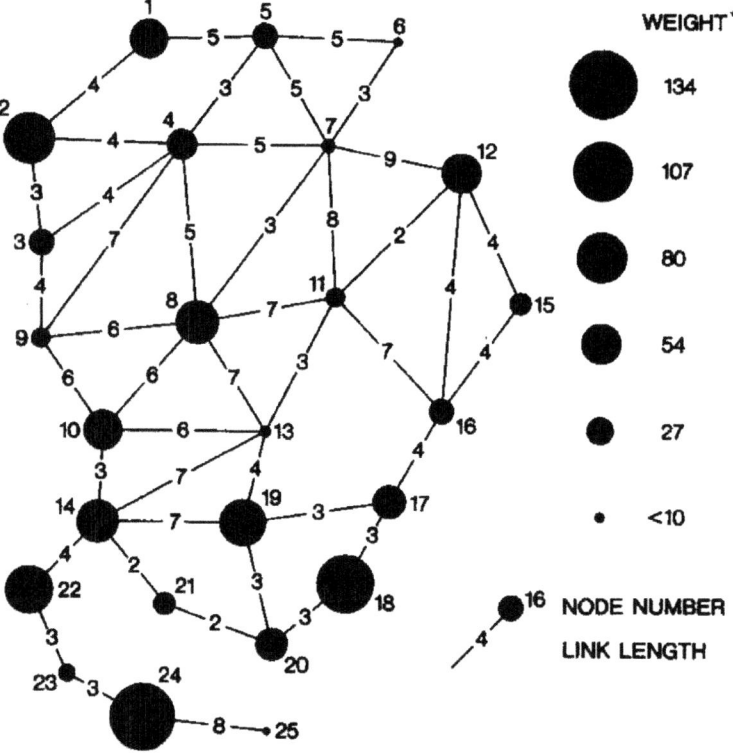

Fig. 18 The regarded test network from [19] is basis for many later studies, such as [21]; nodes are solely used for modeling junctions

5.2 Formulation of the Adapted Model

A model shall be formulated for which algorithms can be applied to determine optimal locations for facilities in a network, where positions on *edges* are regarded as candidates for the location of facilities (as opposed to nodes). Edges are further split into equidistant segments $k \in K$ that are each regarded as potential candidates for the placement of a charging station. The purpose of the algorithm is to determine the segments on which to place facilities to maximize the percentage of vehicles that are charged with (at least) the amount of energy *during their operation* which they consume along their route. A modified model will need to take into account the amount of time that vehicles are standing in the proximity of charging stations by evaluating the cumulated energy along their routes/trajectories and balance this value with the amount of energy that a vehicle consumes along its route. This level of detail cannot be delivered by simply regarding the cumulated flow over a given duration, since many operational aspects (i.e., the precise spatial and temporal distribution and behavior of vehicles) need to be regarded. Next to fully equipping a traffic system and all of its participants with measurement systems, only microscopic traffic simulations can deliver a comparably high level of detail.

In analogy to [19–21], the number of charging stations c_n to be placed is known. Traffic on the other hand and its participants (vehicles) $v \in V$ are attributed with a route, which results in a vehicle trajectory, represented as a list of positions R_v over time, and an amount of energy $e_{v,r}$ that was consumed along that vehicle's trajectory. Each vehicle will pass a number of charging stations along their route, where they will get recharged with the charging power P_{chrg} (net power after subtracting transfer losses) if certain operational conditions are met (e.g., stopped for a time $t > t_{s,min}$ in the required proximity of charging station). $e_{v,k}$ is the amount of energy that was charged into vehicle v at location k by a charging station.

The objective of an optimization algorithm would be to

$$\text{maximize } Z = \sum_{v \in V} y_v, \tag{5.1}$$

i.e., to recharge as many vehicles as possible, subject to

$$\sum_{k \in R_v} e_{v,c} \geq e_{v,r}, \tag{5.2}$$

i.e., a vehicle v is counted as *recharged* only if the sum of recharged energy is greater or equal to the consumed energy along vehicle v's trajectory R_v, with the installation of

$$\sum_{k \in K} x_k = c_n \tag{5.3}$$

charging stations, where

$$e_{v,c} = P_{chrg} \cdot t_{v,c}, \tag{5.4}$$

i.e., the charged energy equals the product of charging power and charging duration, v is a particular vehicle from
V, the set of all vehicles,
y_v is a binary variable, with
$$y_v = \begin{cases} 1 \text{ if } e_{v,c} \geq e_{v,r} \\ 0 \text{ otherwise,} \end{cases}$$
k is a potential facility location from
K, the set of all potential facility locations,
x_k is a binary variable, with
$$x_k = \begin{cases} 1 \text{ if there is a charging station at location } k \\ 0 \text{ otherwise,} \end{cases}$$
R_v is the trajectory of vehicle v, represented as a list of vehicle v's positions over time,

P_{chrg} is the energy transfer rate (charging power) of charging stations,
$t_{v,c}$ is the time a vehicle v gets charged at charging station c along its trajectory R_v,
c is a particular charging station from
C, the set of all charging stations, and
c_n is the number of charging stations to be located.

6 Summary and Outlook

This contribution presented eNetEditor, a newly developed tool that assists users with the rapid prototyping of traffic scenarios for the microscopic traffic simulation tool SUMO. Next to declarations for default SUMO functionalities, eNetEditor was developed to support the definitions of vehicles and traffic demand with the energy model introduced in [6], while keeping it extensible for the support of arbitrary new developments performed in SUMO and any of its extensions.

eNetEditor was developed for city and traffic planners to evaluate the application of alternative energy supply systems in early idea and concept phases, where such detailed vehicle operation data is not readily available and only general traffic measurements exist. It was motivated by the projects *emil* and *InduktivLaden*, both funded by the German Federal Ministry of Transport and Digital Infrastructure. An inductive charging infrastructure and a compatible fully electric prototype bus fleet were successfully integrated in Braunschweig's traffic infrastructure and public transport. With the application of Bombardier's inductive charging system PRIMOVE, vehicles can be charged with powers beyond 200 kW [22]. A crucial factor for the reliable operation of vehicles with limited range is the optimized placement of charging stations. In the project *emil*, infrastructure placement was optimized on the basis of measured public transport vehicle trajectories [23] using the FMS interface [18] and GPS loggers. In project *InduktivLaden*, optimization algorithms shall be utilized in microscopic traffic simulations to allow for the optimization for arbitrary and user-specifiable vehicle and traffic types.

The ultimate goal for the current project *InduktivLaden* is to adapt the presented modifications of flow capture optimization algorithms from chapter "A SUMO Extension for Norm-Based Traffic Control Systems" and to apply them in microscopic traffic simulations. The optimal charging infrastructure locations and their distribution shall be determined for a road traffic system to help in initial estimations of installation costs. While the presented model assumes that the amount of charging stations to be placed is known, the implementations shall include different variants of the optimization algorithms that allow users to define the percentage of vehicles to be adequately recharged along their routes and to optimize the charging infrastructure locations accordingly. Future work further includes the evaluation of the robustness of optimal solutions.

While a prototype version of eNetEditor has been implemented in MATLAB, future work will focus on finishing migration to a platform-independent C++ implementation in *Qt Library* by the end of 2015. Additionally, energy-relevant initialization parameters for vehicle instances will be integrated into routers, along with the presented evaluation routines for optimal arrangement of energy supply infrastructure components for specific fleets and/or overall traffic. Further development will also include the application of sophisticated energy consumption models for energy-optimized routing and calibration, e.g., in cadyts or DUAROUTER.

Regarding the validity of the example scenario presented in chapter "Connecting Macroscopic and Microscopic Traffic Assignment", calibrated vehicle routes will be checked next against the measured public transport vehicle trajectories. Future implementations in eNetEditor itself include the assistance in modeling public transport by the creation of recurring vehicle routes and their integration in calibrated scenarios. For a more realistic integration and application of DFROUTER's and cadyts' functionalities, user-specifiable daily traffic intensity curves as well as lane-based measurement variations and distributions between passenger and commercial/freight vehicles (PKW and LKW) will be implemented. Regarding the application of the newly developed energy device [6] for SUMO, routers are currently not compatible with the definition of vehicles' initial energy content at their departure (in analogy to departSpeed). It is planned to adapt the routers correspondingly or to implement a parser of vehicle route declarations to allow for custom initial vehicle energy content distributions. In the long term, an additional import function for .net.xml files is planned.

References

1. statista. Anzahl der Neuzulassungen von Personenkraftwagen in Deutschland im Jahr 2014 nach Marken. http://de.statista.com/statistik/daten/studie/167008/umfrage/neuzulassungen-von-pkw-nach-marken-in-deutschland/, Rev. 02. February 2015
2. statista. Anzahl der Neuzulassungen von Elektroautos in Deutschland in den Jahren 2003 bis 2014. http://de.statista.com/statistik/daten/studie/244000/umfrage/neuzulassungen-von-elektroautos-in-deutschland/, Rev. 02. February 2015
3. statista. Anzahl der Tankstellen in Deutschland von 1950 bis 2014. http://de.statista.com/statistik/daten/studie/2621/umfrage/anzahl-der-tankstellen-in-deutschland-zeitreihe/, Rev. 02. February 2015
4. BDEW. BDEW-Erhebung Elektromobilität: Zuwachs bei öffentlichen Lademöglichkeiten. https://www.bdew.de/internet.nsf/id/bdew-erhebung-elektromobilitaet-zuwachs-bei-oeffentlichen-lademoeglichkeiten-de, Rev. 02. February 2015
5. Krajzewicz D, Erdmann J, Behrisch M, Bieker L (2012) Recent development and applications of SUMO - Simulation of Urban MObility. Int J Adv Syst Meas 5(3 and 4):128–138
6. Kurczveil T, López PA, Schnieder E (2014) Implementation of an energy model and a charging infrastructure in SUMO. In: Simulation of urban mobility. Lecture notes in computer science, pp 33–43
7. Schnieder L, Stein C, Schielke AG, Pfundmayr M (2011) Effektives Terminologiemanagement als Grundlage methodischer Entwicklung automatisierungstechnischer Systeme. at - automatisierungstechnik 59(1):62–70
8. Schnieder L (2010) Formalisierte Terminologien technischer Systeme und ihrer Zuverlässigkeit. Dissertation, Technische Universität Braunschweig
9. German Aeroscace Center. Simulation/Traffic Lights. http://sumo.dlr.de/wiki/Simulation/Traffic_Lights, Rev. 30. December 2014, 14:24

10. Wegener A, Piorkowski M, Raya M, Hellbrück H, Fischer S, Hubaux J-P (2008) TraCI: an interface for coupling road traffic and network simulators. In: CNS '08 - 11th communications and networking simulation symposium, Ottawa
11. Nguyen T, Fullerton M, Krajzewicz D, Mai ST (2014) DFROUTER - route estimate method based on detector data. In: SUMO2014 - modeling mobility with open data, Berlin
12. Papaleontiou LG (2008) High-level traffic modelling and generation. Master thesis, University of Cyprus, Nikosia
13. Gawron C (1998) Simulation-based traffic assignment - computing user equilibria in large street networks. Ph.D. dissertation, University of Köln, Cologne
14. Flötteröd G, Bierlaire M, Nagel K (2011) Bayesian demand calibration for dynamic traffic simulations. Transp Sci 45(4):541–561
15. Forschungsgesellschaft für Straßen- und Verkehrswesen e.V. Handbuch für die Bemessung von Straßenverkehrsanlagen (HBS). FGSV Verlag GmbH, Köln, 2009
16. Behrisch M, Krajzewicz D, Wang Y-P (2008) Comparing performance and quality of traffic assignment techniques for microscopic road traffic simulations. In: DTA2008 international symposium on dynamic traffic assignment, Leuven
17. Stadt Braunschweig, WVI GmbH. Verkehrsmengen im Werktagsverkehr Mo – Fr in [Kfz/24h] - Querschnittswerte. Stadt Braunschweig, Braunschweig, 2009
18. HDEI/BCEI Working Group (2012) Fleet management system standard description. HDEI/B-CEI Working Group, Version 03
19. Hodgson MJ (1990) A flow capturing location-allocation model. Geogr Anal 22(3):270–279
20. Berman O, Larson RC, Fouska N (1992) Optimal location of discretionary service facilities. Transp Sci 26:201–211
21. Kuby M, Lim S (2005) The flow-refueling location problem for alternative-fuel vehicle. Socio-Econ Plan Sci 39(2):125–145
22. Meins J, Graffam C (2011) Induktive Energieübertragung für Elektrobusse nutzen - Ein Mosaikstein in der Elektromobilität wird in Braunschweig erprobt. *Der Nahverkehr*, 9/2011, 18–20, Alba
23. Kurczveil T, Schnieder E (2014) Measurement evaluations for the arrangement of an inductive charging infrastructure for public transport in Braunschweig (Messdatenauswertung für die Auslegung einer induktiven Ladeinfrastruktur für den öffentlichen Personennahverkehr in Braunschweig). In: AAET 2014 - Automatisierungssysteme. Assistenzsysteme und eingebettete Systeme für Transportmittel, Braunschweig

Printed by Printforce, the Netherlands